CONTROLLING THE BOMB

CONTROLLING THE BOMB
Nuclear Proliferation in the 1980s

LEWIS A. DUNN

A Twentieth Century Fund Report

New Haven and London

Yale University Press

Designed by Nancy Ovedovitz and set in Baskerville type by P & M Typesetting, Inc.
Printed in the United States of America by Vail-Ballou Press, Binghamton, N.Y.

Library of Congress Cataloging in Publication Data

Dunn, Lewis A.
    Controlling the bomb.

    "A Twentieth Century Fund report."
    Includes bibliographical references and index.
    1. Nuclear nonproliferation.   2. Atomic weapons and disarmament.   I. Title.
JX1974.73.D84   327.1'74   81-16086
ISBN 0-300-02820-2   AACR2
ISBN 0-300-02821-0 (pbk.)

10 9 8 7 6 5 4 3 2

For Bobby

CONTENTS

Foreword   ix

Acknowledgments   xi

Introduction   1

1. The First Decades   5

2. The Changing Technical Environment   24

3. How Many Countries Will Get the Bomb?   44

4. What Difference Will It Make?   69

5. Checking the Bomb's Spread   95

6. Building Proliferation Firebreaks   134

7. Mitigating Proliferation's Consequences   149

The Tasks Ahead   176

Notes   181

Index   205

Illustration follows page 99

FOREWORD

The threat of a nuclear holocaust, which first became a reality with the bombing of Hiroshima and Nagasaki, has been contained for thirty-five years. Despite that reassuring record, the risk of nuclear conflict seems likely to rise dramatically in the years ahead as an increasing number of countries—many of them politically unstable—start or resume nuclear weapons programs. American efforts to contain proliferation, initially surprisingly effective, have in recent years focused on the relatively narrow objective of preventing countries from acquiring the capability to make nuclear weapons. Such efforts have become inadequate with the spread of peaceful applications of nuclear energy, which has given rise to both a legitimate market for materials and facilities and to gray and black markets, where technology and resources are freely traded. Moreover, policy designed to restrict the spread of technology cannot deal with the possibility, no longer confined to the pages of science fiction, that a terrorist group might gain access to the bomb. Thus, the risks are multiplied.

Because the threat of nuclear proliferation has long been a topic of concern to the Trustees of the Twentieth Century Fund, they were receptive when Lewis Dunn proposed an examination of American policy and how it might be strengthened. With the Fund's support and under its supervision, Dunn analyzed the pressures that motivate governments to "go nuclear"; he then evaluated the potential for future proliferation, seeking to determine whether it could continue to be slow and limited or whether it would greatly increase and possibly proliferate to runaway proportions, thus enhancing the danger of total annihilation.

Dunn finds that the United States cannot hope to maintain the relatively limited proliferation that had characterized the first generation of nuclear arms, but he does not merely wring his hands or bury his head in the sands at this dangerous new prospect. Instead, he engages in a careful search for the most effective ways to keep proliferation, which appears bound to increase, from getting out of control. Thus, his study proposes a series of measures that the United States might consider to limit the size and the sophistication of the next generation of nuclear forces.

Dunn's study may be viewed as alarmist, but we found him to be a cautious scholar and analyst who believes that problems are best resolved through an awareness of all their aspects— political, economic, social, and technological. Obviously, the possession of nuclear weapons also entails an appreciation of the irrational and emotional in the affairs of nations, and Dunn is at pains to keep the irrational from gaining too powerful a role. His is an intelligent guide to the risks of increasing nuclear proliferation and to possible solutions to them. As such, it is a very worthwhile contribution to both public debate and policy. The Fund is grateful to him for carrying it out.

M. J. Rossant, Director
The Twentieth Century Fund
October 1981

ACKNOWLEDGMENTS

I wish to thank the Twentieth Century Fund for having provided the financial support needed to research and write this
book. Gratitude also is due Hudson Institute, which provided
me with the initial opportunity to begin work on these issues
and with a congenial and supportive environment to complete
the book.

Over the years many persons have helped shape and
sharpen my thinking about the problem of nuclear nonprolif-
eration. Special thanks are extended to the late Donald G.
Brennan and to Herman Kahn at Hudson Institute, to George
Quester of Cornell University, and to Joseph Nye of Harvard
University. I also am indebted to the various individuals with
whom I spoke on trips to the Middle East, Far East, South
Asia, Europe, and South America.

Carol Barker and Marc Plattner of the Twentieth Century
Fund as well as its director, M. J. Rossant, offered comments
and criticisms on earlier drafts of the book. I am grateful for
that aid in improving the logic and precision of its argument.
Pamela Gilfond, my editor at the Fund, worked unstintingly to
enhance the book's readability and punch. Nancy Hoagland,
my assistant at Hudson Institute, also deserves special thanks.

Not least, I want to thank my mother and my wife, without
whose support, encouragement, and understanding this book
would not have been possible.

Since completing the book, I have left Hudson Institute and
accepted a position in the United States Department of State.
I alone, however, am responsible for the arguments, specula-

tions, and proposals that follow. The views expressed do not necessarily reflect those of Hudson Institute or of the United States government or any of its agencies or departments.

INTRODUCTION

The relative stability of the early decades of the nuclear age—
when the United States' monopoly on the bomb was lost but
the spread of nuclear weapons was slow and limited—may be
ending. We are now entering a much more dangerous stage of
proliferation, in which possession of the bomb by countries in
conflict-prone regions is not only possible but probable and the
threat of actual use of nuclear weapons is growing.

The United States' success in building an atomic bomb in
1945 triggered a proliferation chain that encompassed four
countries—the Soviet Union, Britain, France, and China—over
nineteen years. The conditions that checked the scope and
pace of that proliferation have eroded. For many nations, the
technical barriers to acquisition of nuclear weapons are easier
to overcome. The disincentives to "going nuclear" are becoming less compelling and often are only dimly perceived. Besides, they can be outweighed by heightened security pressures, by growing desires for political status and influence, or
by a belief in the domestic political gains of possessing the
bomb. Moreover, there are foreseeable international events
that may even further undermine the checks on proliferation.
Most important, in almost every region of the world there is at
least one country, and possibly more, seeking to acquire nuclear weapons or likely to do so in the event of readily identi-
fiable developments.

Nuclear weapons proliferation, under way for some thirty-
five years, shows no signs of abating. On the contrary, the pace
of proliferation may quicken, and its scope could be considerable. Widespread proliferation, however, is not inevitable. Much
depends on the policies pursued by the United States and

other countries. While it is unrealistic to expect that additional
nuclear tests can be prevented, it may yet be possible to hold
the line at no more than a continuation of the slow and limited
proliferation pattern of the first decades of the nuclear age.
That is, over the next decade or two only another four or five
countries—perhaps fewer—would covertly or overtly acquire
nuclear weapons.

Contrary to what some people feared in 1945, nuclear weapons have not been used since Nagasaki. But it cannot be
assumed that because the nuclear powers avoided their use
until now, the countries that will be acquiring nuclear weapons
also will refrain from using them. At least some of the conditions underlying the first decades' nuclear peace will be lacking
when nuclear weapons spread to conflict-prone regions.
The result will be a much nastier and dangerous world for the
United States, the other great powers, and particularly the
countries of those regions. And if there is an extensive erosion
of the first decades' pattern, these adverse consequences will be
greater still.

Traditional nonproliferation policies designed to slow the
pace and contain the scope of proliferation must remain,
therefore, at the core of U.S. strategy. Particularly in light of
backsliding on the part of some major nuclear suppliers, exports controls must not only be maintained but must also be
tightened. Improved intelligence about countries' nuclear
weapons activities also could have a high payoff. Even more
important, greater attention must be given to defusing emerging proliferation hot spots. Steps must be taken soon to reduce
incentives to "go nuclear" by enhancing the security of critical
countries, minimizing uncertainties among rival countries
about one another's nuclear weapons intentions, and launching
diplomatic initiatives to alleviate underlying disputes and
sources of conflict. Disincentives to nuclear weapons acquisition also must be heightened by a more credible threat of multilateral sanctions. Moreover, responses must be readied for
handling the dramatic proliferation events of the 1980s, not
least of all, additional nuclear tests. As well, measures must be

devised now for mitigating the regional and global consequences of more widespread proliferation.

Because of the heightened danger of the next decades, there
is a need for continuing, steady bureaucratic attention to and
appraisal of progress in this area. Periodic flurries of intense
concern followed by years of benign neglect will no longer suffice. Further, the limits of U.S. military and political influence
abroad, the rise of new centers of economic power and technical expertise, the end of American dominance over nuclear
energy, and the growing international Soviet role all increase
the importance of international cooperation. But the need to
buttress continuing unilateral American measures with multilateral initiatives adds to the complexities of proliferation
policy.

There inevitably will be clashes between proliferation policy
and other significant U.S. foreign, domestic, economic, and security policies. For example, steps to enhance the security of a
nonnuclear weapons state may conflict with domestic reluctance to take on an active security role in the Third World; the
imposition of sanctions to support global nonproliferation
goals may have adverse domestic economic repercussions and
clash with other American objectives in the region. Nonproliferation policy will not, of course, always be at odds with
broader U.S. foreign policy. In Southwest Asia, for example,
the likelihood of overt acquisition of nuclear weapons by Pakistan might be reduced by closer U.S. security ties to that country as part of a broader American role in the Persian Gulf.

Nevertheless, the adverse impact of increased proliferation
on a broad range of U.S. interests, as well as the serious limitations of a strategy for coping with proliferation's consequences, strongly suggests paying significant costs in support of
nonproliferation policies. Those who too readily favor a "more
realistic" emphasis on living with proliferation should remember that once nuclear weapons have spread to one or more of
the many conflict-prone regions, we will be faced with the
threat of small-power nuclear wars and the heightened prospect of terrorist control of nuclear weapons.

Above all, it should not be forgotten that no matter what strategies are pursued, it may be impossible to prevent the next use of nuclear weapons. In an anarchic international system, with nations divided by ideology, history, tradition, culture, and geopolitics, there is no certain escape from the threat of nuclear warfare. An understanding of how it all began is the first step to anticipating what may lie ahead and how best to prepare for it.

1 • THE FIRST DECADES

ALONG THE NUCLEAR ROAD

The First Nuclear Weapons State

In October 1939, President Franklin D. Roosevelt authorized feasibility studies and research on the development of a decisive weapon based on atomic fission. It was the first step toward building an atomic bomb and ushering the world into the nuclear age. By the winter of 1942, the Manhattan Project had been established—an all-out commitment to the development of a fission bomb in time for its use in World War II.

These initial American efforts were spurred on by the fear that Nazi Germany was attempting to build an atomic bomb.* American officials were greatly encouraged by two developments in the summer of 1941—Britain's MAUD Committee Report,† which argued that a uranium fission bomb could be

*Until late in World War II, the few American officials cognizant of the Manhattan Project, as well as the scientists and engineers involved, mistakenly thought themselves in a race with German scientists to get the bomb first. Richard G. Hewlett and Oscar E. Anderson, Jr., *The New World, 1939–1946*, vol. 1: *A History of the United States Atomic Energy Commission* (University Park, Pa.: The Pennsylvania State University Press, 1962), p. 17.

†Comprised of about a dozen scientists, the MAUD Committee was organized in April 1940. It addressed two questions: Was a uranium bomb feasible? If so, could Britain develop the bomb before Germany? Answering both questions in the affirmative, the MAUD Committee urged construction of the bomb in its report to the Ministry of Aircraft Production, which had sponsored its work. See Andrew Pierre, *Nuclear Politics: The British Experience with an Independent Strategic Force 1939–1970* (London: Oxford University Press, 1972), pp. 15–20; Margaret Gowing, *Britain and Atomic Energy, 1939–1945* (London: Macmillan & Co. Ltd., 1964), pp. 45–89.

produced in time to affect the war's outcome, and technical information supplied by M. L. Oliphant, an Australian physicist who had served on that Committee.[1]

Following Germany's defeat without use of this "decisive" weapon, the United States rushed to complete the bomb in time to affect the war in the Pacific, where it was hoped, in the words of Secretary of War Henry L. Stimson, that the bomb would administer a "tremendous shock" sufficient to force Japan's surrender and preclude the need for invasion.[2] On July 16, 1945, the United States tested the first atomic bomb fifty miles northwest of Alamogordo, New Mexico. On August 6, 1945, the United States detonated the first atomic bomb used in warfare on Hiroshima. Three days later, a second atomic bomb fell on Nagasaki, bringing World War II to a close.

The United States alone among the first nuclear powers developed the atomic bomb as an instrument of war, with every intention of using it once available.[3] But American policymakers saw as well the political advantages of being the sole possessor of the awesome new weapon and hoped, in particular, that the bomb would strengthen the bargaining position of the United States vis-à-vis the Soviet Union.[4] This belief in the political benefits provided by the bomb would play a role in other countries' decisions to "go nuclear"—decisions that followed in a slow chain reaction from that fateful day in 1939 when the United States set out on the road to Alamogordo.

The Soviet Union Catches Up

The Soviet Union was next. As in the United States and Britain, the discovery of atomic fission stimulated studies of the possibility of both controlled and explosive uranium chain reactions as early as 1939. But these studies were halted by the Nazi invasion in June 1941. Research resumed in 1943, but it was not until 1945, when the successful Alamogordo test and use against Japan erased any remaining Soviet skepticism about the technical feasibility and military value of an atomic bomb, that the Soviet Union established a full-scale research, development, and production program.[5] In July 1946, the Soviet government rejected the U.S.-sponsored Baruch Plan, which proposed to replace by stages the American nuclear monopoly with international control of atomic energy and the bomb.[6] Three years later, in 1949, the Soviet Union detonated its first atomic bomb.

Joseph Stalin clearly recognized that atomic weapons constituted a decisive element of military power, one that prudence dictated the Soviet Union had to have.* Given Stalin's fears of foreign encirclement and the deeply rooted perception of the whole Soviet leadership of the outside world as hostile and threatening, not to have pursued such military power would have been unthinkable.[7] As well, the political disadvantages of allowing the United States to hold a nuclear monopoly were understood. Matching the American bomb became a way of asserting and justifying a Soviet claim to equality with the West, not only to outsiders but to the Soviets themselves.[8]

While the Soviets' doubts about the feasibility of the atomic bomb caused them to hold back on their early research and development efforts, after August 1945 there was no question that nuclear weapons could be made and did work. Those countries now considering "going nuclear" will not be slowed by that once critical technical constraint. And the speed with which the Soviet Union produced an atomic bomb itself cautions against overoptimistic assessments of how long it will take many of these countries to produce atomic weapons.

Britain Follows Suit

Although British nuclear weapons feasibility studies were begun in 1939 and intensified in September 1941, following the submission of the influential MAUD Committee Report, wartime constraints on resources precluded an independent British atomic bomb project. Aside from assistance provided by individual British scientists working on the Manhattan Project,

---

*Milovan Djilas, then a deputy to Tito, reported a conversation he had with Stalin in January 1948:

> Stalin spoke up about the atom bomb: "That is a powerful thing, pow-er-ful!" His expression was full of admiration, so that one was given to understand that he would not rest until he, too, had the "powerful thing."

Milovan Djilas, *Conversations with Stalin* (New York: Harcourt, Brace & World, 1962), p. 153.

Britain's efforts were confined to modest preliminary research. After the war, a full-scale British nuclear weapons program emerged, in part stimulated by the successful conclusion of the Manhattan Project. In 1945, Britain decided to build the facilities necessary to produce plutonium and, in January 1947, to produce an atomic bomb. Less than six years later, in October 1952, Britain became the third country to detonate an atomic bomb.[9]

Britain was motivated to a considerable degree by the simple assumption of its leading politicians and civil servants that to remain a first-rank power it had no choice but to follow the American lead. Disputes with the United States over nuclear collaboration and postwar diplomatic strategy confirmed that assumption.[10] Although Britain believed that possession of nuclear weapons was a useful hedge against an American return to isolationism,[11] more direct military security concerns played little, if any, role in the critical decisions of 1945–47. Those decisions also were encouraged by the scientific and bureaucratic momentum generated by early British research. In 1947, when the decision to proceed was made, Britain had been engaged in research for more than eight years. The development of an atomic bomb was seen simply as the completion of a project delayed by wartime demands.[12]

Political status, global influence, and scientific and bureaucratic momentum were all significant incentives for Britain to "go nuclear." Conspicuously absent was a careful, rational, longer-term assessment of the costs and benefits. But Britain was not the only country to become a nuclear power without paying much attention to the liabilities, difficulties, complications, and mounting costs that lay ahead.[13]

Britain's decision in turn stimulated and reinforced France's incentives for acquiring nuclear weapons. But far more extensively than in Britain, bureaucratic momentum drove France's step-by-step advance to the bomb.

France: One Step behind Britain

France's decision in 1951 to build nuclear reactors especially designed to produce plutonium was the initial step in what came to be a nuclear weapons program. Feasibility studies and preparations for building nuclear weapons were initiated by the Commissariat à l'energie atomique (CEA) in 1954. Less than one year later, a secret protocol was signed by the CEA and the Ministry of Defense (MDN), in which the MDN promised financial support for CEA activities, while the CEA agreed to furnish plutonium to the MDN and to conduct additional preparatory studies. Continuing this incremental advance, the Bureau d'études générales was established in May 1955 within the CEA and given responsibility for designing an atomic bomb. Eighteen months later, in November 1956, a secret four-year military program was begun, with its goal the production of a prototype atomic bomb by the CEA.[14]

Because of the institutional weakness and instability of the Fourth Republic and the lack of parliamentary awareness or interest, by 1956 a nuclear weapons program had emerged in France without an official cabinet-level decision to develop those weapons. Such a decision was not taken until April 1958. By that time, the earlier incremental decisions made through informal channels by a few well-placed individuals had made it little more than a formality.

These early decisions were critically influenced by the desire to maintain equal status with Britain and to reassert France's international prestige. Acquisition of nuclear weapons was seen to symbolize both France's refusal to accept second-rank status and its claim to influence within the Atlantic Alliance. Keeping an eye on British activities, proponents argued that if nuclear weapons were necessary and appropriate for Britain, they were appropriate for France.[15] As with Britain, security-related incentives, including questions about the reliability of the American nuclear umbrella in an era of growing U.S. vulnerability to Soviet atomic weaponry, carried less weight.[16]

With General Charles de Gaulle's return to power and the creation of the Fifth Republic in May 1958, nuclear weapons became part of a more ambitious attempt to create a French-led Europe equal in power and in influence to the superpowers, thus seeking to alter the very structure of postwar world politics. De Gaulle saw the acquisition of nuclear weapons as a

necessary means of restoring the self-respect of an officer
corps torn by defeat in Algeria and Indochina.[17] And at least
for a time in the mid-1960s, a French capability to threaten Soviet cities with nuclear destruction was thought by some influential Frenchmen to be a way to escape from a major European war.[18]

France's entry into the nuclear weapons club with the detonation of an atomic bomb in 1960 typified the process of proliferation during the first decades of the nuclear age. As in
Britain, it was a further link in the proliferation chain reaction
triggered by the United States and more strongly motivated by
the perceived political benefits of "going nuclear" than by direct military security concerns. And in light of the present political and institutional weaknesses of many potential nuclear
weapons countries, France's incremental, step-by-step decision-
making process—in which a handful of scientists and officials
set in motion, nurtured, and brought to fruition a nuclear
weapons program—may be an indicator of things to come.
Also, as in Britain and to some degree even in the Soviet
Union—and quite possibly as it will be in other countries in
the future—there was little assessment of the longer term costs,
military requirements, or complications eventually to be en-
countered in developing a nuclear force.[19]

China Joins the Club

China's efforts to acquire the scientific, technological, and industrial base for a nuclear weapons production program were
under way by the early 1950s. In the mid-1950s, with the
promise of technical assistance from the Soviet Union, a decision was made to establish a nuclear weapons program. The
Soviet Union apparently agreed to supply China with a sample
atomic bomb and data on its design and manufacture.[20] While
the Soviet Union never followed through on this commitment,
it did help China design and construct a gaseous diffusion uranium enrichment plant, which eventually supplied the weapons-
grade uranium for China's first atomic bomb in 1964.

Fear of the United States was originally the major incentive
for China's nuclear weapons program, considered necessary to

deter American nuclear blackmail or even actual use against China. Engendered by American nuclear threats in the closing stages of the Korean War, that fear was reinforced in the late 1950s by the debate within the United States about limited nuclear war and the enunciation by the Kennedy Administration in the early 1960s of the nuclear doctrine of flexible response.[21] Both seemed to threaten the use of nuclear weapons to compensate for the inferiority of American conventional forces in Asia. During the late 1960s and early 1970s, as the Sino-Soviet split intensified, deterring a Soviet nuclear attack increasingly became the predominant motivating force behind China's nuclear weapons program.[22]

Nevertheless, perceptions of the political benefits to be attained by acquiring nuclear weapons also motivated China, as they had other early nuclear powers. Like Britain and France, China regarded nuclear weapons as a necessary buttress to and expression of Chinese sovereignty, political independence, and major-power status. And just as early French thinking about the atomic bomb was influenced and legitimized by the British example, so the French example strengthened the hand of those Chinese advocating "going nuclear."[23]

With China's entry into the nuclear club, the first proliferation chain had run its course. Although China's actions helped push India further along the road to the bomb, India's detonation of a nuclear explosive device in May 1974 was motivated by many other reasons as well. Rather than being a continuation of the past, India's test was a major turning point closing out the first decades of the nuclear age and initiating the bomb's spread to the Third World.

Nuclear Abstinence

During the first decades, some countries technically capable of manufacturing nuclear weapons chose not to do so. Canada, the Federal Republic of Germany, and Japan were prominent among them.*

---

*The smaller countries of Western Europe, Italy, and Australia also were among this group.

Although Canada cooperated with the United States on the Manhattan Project, it decided in 1946 not to pursue the military uses of atomic energy. The fact that Canada's security was credibly assured by the United States weighed heavily in this initial decision. Later it also was thought that development of nuclear weapons might encourage proliferation, thus undermining Canada's support for global disarmament measures. Besides, acquisition of nuclear weapons seemed of little relevance for realizing Canada's goal of closer economic and political ties with the newly independent countries of Asia and Africa. Once the decision crystallized, there was no reason to reassess Canada's initial choice.[24]

The Federal Republic of Germany, which also had the necessary technical and industrial capability, renounced in the 1954 Paris Treaties the right to manufacture atomic weapons on its territory.[25] Underlying its continued nuclear abstinence has been the military security provided by the Atlantic Alliance, within which the Federal Republic also was able to influence American strategy. Not only were there no compelling reasons to "go nuclear," there were some decided political, if not military, risks associated with doing so—including worsening relations with the Soviet Union, disruption of the North Atlantic Treaty Organization (NATO) security structure, a return to being Europe's outcast, and probable domestic opposition.[26]

Since World War II, Japan's policy has been guided by the "three nonnuclear principles": neither to acquire nuclear weapons, manufacture them, nor allow their introduction into the country by a foreign power. Because of Japan's so-called nuclear allergy—the result of being the target of atomic bombing—domestic political opposition to developing nuclear weapons has been strong. Compelling security, status, or influence-related incentives also have been lacking: the United States–Japan Mutual Security Pact of 1960 has provided for Japan's security, and Japan's growing economic power has been a far more effective source of regional and global political influence. In fact, many in Japan have argued that "going nuclear" would lead to greater vulnerability, reduced security, and lessened international influence.[27]

Second Thoughts on the Nuclear Road

It has been said that "the decision to develop nuclear weapons is not a fluke of certain governments but a general technological imperative."[28] As the momentum generated by scientists in Britain indicates, this contention is not wholly unfounded. But the strength of technological momentum must not be exaggerated in an evaluation of the prospects of further proliferation. For not only did some countries technologically capable of producing an atomic bomb in the first decades eschew such a possibility, several others—among them, Sweden and Switzerland—abandoned feasibility studies and preliminary preparations for producing nuclear weapons.

While the potential implications of atomic weaponry for Sweden's defense had from the very beginning concerned some members of its defense establishment, the question of acquisition of nuclear weapons was publicly broached for the first time in 1954 by the Swedish Army commander-in-chief. The military, influenced by NATO's switch to reliance on battlefield nuclear weapons and by British White Papers emphasizing the importance of nuclear weapons for defense, claimed that nuclear weapons were needed to buttress Sweden's defenses, thereby making violation of its neutrality too costly. Discussion ensued, but, in 1959, the governing Social Democratic Party postponed decision lest it split the party and lead to its fall from power. In the following years, this shift in opinion away from acquiring nuclear weapons was reinforced by studies conducted by the Swedish Ministry of Defense. These studies—uncharacteristically analytic compared with other nuclear decision making of the first three decades—concluded that, given a fixed level of defense spending, acquiring battlefield nuclear weapons at the cost of sacrificing more useful conventional defenses would be a mistake.[29] By the mid-1960s, nuclear weapons acquisition was no longer an issue.

Serious debate and preliminary nuclear weapons preparations were also begun in Switzerland in the mid- to late-1950s, stimulated by NATO's growing emphasis on the battlefield use of nuclear weapons. At issue was the utility of battlefield nuclear weapons for deterring invasion; except for contentions that the Swiss tradition of armed neutrality demanded the "most modern" weapons, prestige and status played no role. As in Sweden, perceptions of the high costs involved led Switzerland in the mid-1960s to shy away from a commitment to acquire nuclear weapons.[30]

There are also indications that secret, low-level nuclear weapons preparations were begun by Yugoslavia in the late 1950s or early 1960s. President Tito became convinced, however, that scarce defense resources could be spent more wisely elsewhere, and those preliminary activities were halted.[31]

Four Proliferation Failures

In 1940, Japan initiated preliminary studies on building an atomic bomb. But at a series of ten fateful meetings of Japan's leading scientists between December 1942 and March 1943 it was concluded that Japan could not build a bomb in fewer than ten years. Nor, so the scientists argued, could the United States build an atomic bomb in time for its use in World War II. Consequently, although nuclear research did continue in Japan until the end of the war, it was sporadic, disorganized, and without substantial funding. While Japan did not succeed in producing an atomic bomb, it should be remembered that the uncertainty about the prospects for success that led to at most halfhearted official support no longer exists. And even then, some advances were made—the product of efforts by individual scientists and the momentum their work generated.[32]

Nazi Germany's efforts to produce an atomic bomb began as early as August 1939.[33] But the project failed. An erroneous scientific calculation by a physicist whose word could not be questioned because of his high status led the Germans to concentrate on the time-consuming and more technically demanding development of heavy water–moderated plutonium production reactors rather than on the graphite-moderated reactors that produced the plutonium used in the Alamogordo and

Nagasaki bombs. Rivalries among groups of scientists also hindered the early research. But of greater significance, and exemplifying once again the limited rationality of past nuclear policymaking in general, the peculiar bureaucratic environment of wartime Nazi Germany contributed to the failure. Pessimistic estimates about how long it would take to make a bomb by German physicists afraid of risking failure were one reason why the government was unwilling to commit itself to more than preliminary preparations. Hitler's lack of interest in atomic energy was also an inhibiting factor.

But even with limited official support, German nuclear scientists, like their Japanese counterparts, continued their research until the end of the war. A project thought "scuttled" in 1942 by then Minister of Armaments Albert Speer was in fact carried forward by scientific curiosity and momentum.[34] When the war ended, work was under way on theoretical nuclear physics, the design and operation of a nuclear reactor, and uranium enrichment.[35]

Unlike Japan or Nazi Germany, Argentina and Libya attempted to buy their way into the nuclear club in the first decades. Soon after the creation of the Argentine National Atomic Energy Commission in 1950, President Juan Perón employed Austrian emigré and nuclear scientist Ronald Richter. Rumors of a forthcoming Argentine Manhattan Project and bomb quickly began to circulate in Buenos Aires. Two years later, however, with little apparent evidence of progress, Richter was fired, thus ending Perón's attempt to build an atomic bomb.[36] As for Libya, Colonel Muammar Qaddafi is reported to have attempted unsuccessfully in 1971 to purchase nuclear weapons from the People's Republic of China.[37]

THE CAUSES OF NUCLEAR RESTRAINT

Technical Constraints

When the atomic bomb fell on Hiroshima, one key technical constraint—uncertainty about the feasibility of such a bomb—ceased to exist. Nevertheless, in the first decades of the nuclear age, technical constraints were significant in checking prolifer-

ation's scope and pace. Manufacturing nuclear weapons entailed overcoming an array of theoretical, technical, scientific,
engineering, industrial, and organizational problems. It is true
that for some countries, such as the Federal Republic of Germany, Canada, or Japan, these technical constraints played little role in their decisions not to "go nuclear." While for others,
technical constraints only slowed the pace of their nuclear
weapons programs. But for most developing countries without
sufficient indigenous industrial and technical development or
access to competent outside assistance and materials—represented, for example, by Argentina in 1950 or Libya in 1971—
these technical constraints were a boundary that could not be
crossed.

Limited Incentives

The lack of highly compelling incentives for many countries to
"go nuclear" was of even greater importance in limiting the
scope and pace of proliferation in the first decades. While enhanced military and political security was a major stimulus for
several of the first nuclear powers, most other countries concluded that acquisition of nuclear weapons was not necessary
to ensure their political independence, territorial integrity, or
survival. American alliances in Europe and Asia, adequate conventional defense postures, and the lack of any hostile nuclear-
armed regional rivals all contributed to the perception that
other sources of security were sufficient.

Each of the original nuclear powers at least to some degree
equated status and influence with acquisition of the bomb.
Nevertheless, possession of nuclear weapons did not come to
be widely regarded as a stepping stone to greater international
prestige and influence by middle-rank powers, let alone by developing nations. The recovery of Japan and West Germany
following their defeats in World War II demonstrated that
there were other paths to international standing. In fact, with
the growing acceptance toward the close of the first decades of
a norm of nonproliferation, there began to emerge a negative
connotation to acquiring nuclear weapons.

Further, "going nuclear" was not commonly seen to have a

significant domestic political payoff. While de Gaulle saw acquisition of nuclear weapons as a means of strengthening bureaucratic and public morale, few leaders shared his point of view. Nor, with a few exceptions, was scientific and bureaucratic momentum a powerful driving force.

As well, the small number of countries that did "go nuclear" and the slowness of the pace of nuclear weapons acquisition contributed to the limited growth of incentives. Acquisition of these weapons remained the exception to the norm, not a model to be followed lest a country lose its self-respect and status or because global proliferation was inevitable.

Disincentives to "Going Nuclear"

Compelling disincentives to acquiring nuclear weapons also underlay the slow and limited pattern of proliferation in the first decades of the nuclear age. Nuclear decision making in West Germany, for example, was clearly affected by fear of adverse foreign reaction. Japan feared domestic opposition. Concern that nuclear weapons would be pursued at the unacceptable cost of sacrificing necessary conventional defense capabilities played an important role in Sweden's nuclear policy debate.

Reinforcing these disincentives was a growing norm of nonproliferation, a perception of the questionable legitimacy of acquiring nuclear weapons. The different ways in which France in 1960 and India in 1974 crossed the nuclear explosive threshold symbolize this change in attitude. France acknowledged its intention to test a nuclear weapon and proclaimed its emergence as a nuclear power. India, to the contrary, prepared in secret and then detonated what its leaders termed a "peaceful nuclear explosive."

The Treaty on the Nonproliferation of Nuclear Weapons (NPT) perhaps better illustrates this norm of nonproliferation. Negotiated primarily by the United States and the Soviet Union during the mid-1960s, the NPT was opened for signature in 1968. By the mid-1970s, more than one hundred countries had agreed under its provisions not to manufacture or develop nuclear weapons and had accepted safeguards—ranging from international inspection of peaceful nuclear activities to supervised accounting of all materials used in such activities—designed to detect any illegal diversion of material or facilities for military purposes. In exchange, the NPT recognized the rights of parties to the treaty to develop peaceful nuclear energy and pledged the assistance of the advanced countries for that purpose. The NPT itself is a significant disincentive to acquisition of nuclear weapons, for although it provides for a right of withdrawal, a country wishing to do so must overcome possibly strong public and bureaucratic reluctance to reverse past policy and publicly reject the internationally accepted norm of nonproliferation.

There is one caveat qualifying the impact of these disincentives. Where the initial incentives for acquiring nuclear weapons were sufficiently powerful, the disincentives to and complications of acquiring the atomic bomb appear not to have been thoroughly analyzed or taken into account. Britain, France, the Soviet Union, and even the United States paid only limited attention to the manifold costs, ramifications, and political and military requirements that would accompany nuclear weapons acquisition.

NUCLEAR PEACE SINCE NAGASAKI

Contrary to the fears of many people at the time, nuclear weapons have not been used since August 9, 1945, when the United States dropped an atomic bomb on Nagasaki. Nuclear peace was not, however, an automatic by-product of those weapons' awesome destructiveness. It was the result of particular geopolitical and technical features of the Soviet-American nuclear confrontation—backstopped by more than a little luck.

The Geopolitics of Nonuse

The stakes of the Soviet-American confrontation remained limited—even at the height of the Cold War. Neither country sought to challenge the territorial integrity or political independence, let alone the physical survival, of the other. Despite periodic East-West crises from the late 1940s to the early 1960s over Western Europe's political orientation, neither superpower believed that recourse to military force and the risk of escalation to a nuclear exchange were justified. The Eisenhower Administration's talk of "rolling back" the Soviets in Eastern Europe remained talk; the Soviets stopped far short of direct military action in their efforts to change the status quo in Berlin. Outside of Europe, much of the Soviet-American confrontation was indirect, stemming from respective efforts to exercise global influence, support friendly governments, and preserve their reputations for willingness to use force in support of their vital interests. But such goals did not reach a level of importance that might warrant serious thinking about recourse to nuclear weapons.*

In each direct Soviet-American confrontation, moreover, the "balance of interests" clearly favored one side.[38] In the crisis over Western access to Berlin in the late 1950s or the Cuban Missile Crisis of 1962, for example, U.S. interests outweighed—and eventually were seen by the Soviets to outweigh—Soviet interests in changing the status quo. Consequently, it was easier to resolve these crises peacefully than it would have been if both sides believed that the only legitimate outcome was for the other to give way.

That the Soviet Union and the United States are not geographically contiguous also contributed to the first decades' nuclear peace. Distance meant that quite a few steps had to be taken before a Soviet-American crisis or confrontation escalated to nuclear attacks against their own societies. Consequently, there existed more room for maneuver and a greater margin for error, thereby lessening the danger of hasty decisions, miscalculations, or overreaction out of a fear that even a local setback might rapidly call survival into question. In addition, through the Berlin Airlift Crisis of 1948, the Korean War in 1950, and the Berlin Crises of 1958 and 1961, each side ac-

---

*On several occasions the United States apparently gave passing thought to using nuclear weapons, but *not* where the Soviets were involved. For example, during the French defeat at Dien Bien Phu, some members of the Eisenhower Administration—but not President Eisenhower—thought of using tactical nuclear weapons to break the siege. See George Quester, *Nuclear Diplomacy, The First Twenty-Five Years* (New York: Dunellen Publishing, 1970), p. 124.

quired a sense of how the other might act and react under pressure, accumulated experience in crisis diplomacy, and came to recognize certain rules of the game. Those rules included, for example, avoiding a direct armed clash with the other, continuing communication during crises, limiting possibly unmanageable bellicose demonstrations, and not changing the status quo by sudden military faits accomplis.[39] The deployment of missiles in Cuba, which violated this last rule, represented a major Soviet miscalculation of what the United States would accept as within the legitimate boundaries of confrontation. Even so, that crisis occurred after nearly fifteen years of adjusting to living with each other's nuclear weapons, which may have contributed to its successful resolution. That extended learning process was not unrelated to the greater leeway and flexibility provided by the fact that the protagonists' homelands were outside the initial crisis zone.

Also among the reasons for the passage of the first decades without further use of nuclear weapons was the presence of prudent and cautious leaders. They carefully weighed the consequences of given actions and did not permit hatred or contempt for their opponent to distort their policies. With the one major exception of Khrushchev's decision to place missiles in Cuba in 1962, the leaders of the two superpowers also shied away from potentially reckless or adventurist gambles. And once Khrushchev recognized his miscalculation, he sought peaceful resolution of the crisis. President John Kennedy, for his part, provided Khrushchev with a face-saving way out and resisted any temptation to humiliate him.

Also critical to nonuse was the emergence of the nuclear taboo holding that those weapons were not merely more advanced conventional ones. Contributing to its creation was the fact that the atomic bomb was not used in the Korean War. Among the reasons for that restraint were a reluctance to use atomic bombs against Asians again, British pressure, and difficulties in figuring out how to use the bomb in that particular case. But especially important was the military's belief that its limited stockpile of atomic bombs had to be preserved for use against the Soviets in what was feared would be an imminent war in Europe.[40]

Finally, directly fostering the caution of the Soviet and American leaders, and thereby the nonuse of nuclear weapons, was the grave destructiveness of the atomic bomb. Many analysts thought that war and politics had been severed, that the cost of nuclear war outweighed any potential purposes that it might serve and therefore that both the United States and the Soviet Union had to avoid actions that could lead to such a war.[41] But though of considerable importance, the inhibiting effect of nuclear weapons themselves may not suffice to explain nonuse in the first decades. Had the stakes been thought sufficiently high, or had less prudent leaders been present, or had the breathing space provided by geography not been there to hinder escalation, the nuclear peace could have been shattered—as many people originally feared in 1945.

The Minimal Technical Requirements

Certain technical capabilities also contributed to the nonuse of nuclear weapons in the first decades. Their presence made it unnecessary for leaders to act quickly and with only limited information, reduced pressures to escalate in an intense crisis, and greatly lessened the chance of unsettling shocks that could lead to a nuclear conflagration.*

Most American analysts would agree that one minimal technical requirement of stable deterrence between the United States and the Soviet Union has been both sides' possession of a second-strike capability—the capability to survive a surprise first strike against one's own nuclear force and then to be able to retaliate, inflicting "unacceptable" damage on the attacker.[42] If the United States or some new nuclear power lacked that capability, a desperate opponent might conclude that a nuclear strike was its least undesirable option in an intense crisis or escalating conventional conflict. Equally important, neither the United States nor the Soviet Union relied on a hair-trigger, launch-on-warning procedure to protect its strategic forces against surprise attack. They thereby greatly lessened the risk

---

*The requirements of stable deterrence have been increasingly debated in recent years. What follows attempts, at the risk of some oversimplification, to set out only some of the generally accepted minimal requirements.

of an accidental war due to mechanical or human failure in a tense crisis.[43]

Measures to reduce the chances of an accidental detonation of a nuclear weapon also buttressed the first decades' nuclear peace. In an intense crisis, let alone with conflict under way, an accidental detonation—perhaps caused by the crash of an airplane with nuclear weapons on board—could spark a larger conflict. Similarly, steps to maintain tight control over nuclear weapons and ensure that they could not be used without authorization by the highest leaders also were necessary. Preventing an accidental or unauthorized use of nuclear weapons has been part of the overall development of a stable and low-risk peacetime nuclear posture. For example, the adoption of "failsafe" procedures, requiring a bomber placed on airborne alert not to go beyond a given distance to its target without a positive signal to do so, was one such step in this direction.

Nonetheless, these measures that reduce the propensity to war often were developed and adopted later than might have been expected. For example, in the early 1950s, the concept of a second-strike capability was only beginning to be assimilated and put into operation by the United States military. And until the early 1960s, the strategic balance was such that the United States could have used nuclear weapons first against the Soviets and suffered far less damage from a retaliatory strike than would the Soviets should they have inflicted a first strike. Similarly, although considerable intellectual and financial resources were devoted to designing against accidental detonation of a nuclear device from the early 1950s onward,[44] a concerted American effort to assure against unauthorized access and use was not undertaken until the late 1950s and did not receive political support until the early 1960s.* Even so, Soviet command-and-control concepts lagged behind, despite unofficial but authorized American-Soviet communication on this

---

*According to professionals dealing with these matters, not until the early 1960s was there a political interest in techniques and procedures that the weapons laboratories could have implemented earlier. Also see Joel Larus, *Nuclear Weapons Safety and the Common Defense* (Columbus: Ohio State University Press, 1967), pp. 80–86.

matter.[45] And the original meaning of "fail-safe" in the United States in the late 1940s and early 1950s was to proceed if no communication *not* to do so had been issued.[46]

Meeting these minimal technical requirements of nuclear deterrence has been a time-consuming, technically demanding, and costly task. Nor can it be said even now that the task is completed.[47] But while the geopolitics of the U.S.-Soviet conflict compensated for various technical shortcomings in the first decades of the nuclear age, a comparable compensating effect may be lacking in the conflict-prone regions to which nuclear weapons are likely to spread.

Chance

Chance has also contributed to the nonuse of nuclear weapons in the first decades. It alone accounts for the fact that the first Soviet ship to reach the naval blockade during the 1962 Cuban Missile Crisis was an oil tanker that did not have to be searched because it could not have been carrying additional missiles, thus providing more time for Khrushchev to decide how to answer Kennedy's ultimatum. More important, chance was reflected in the fact that atomic weapons were not used by the United States in November 1950 after the Chinese army intervened in the Korean War, badly mauling American forces and nearly pushing them into the sea. Had the very limited United States stockpile of nuclear weapons been larger, or had some of them been deployed in South Korea, pressures to use them to avoid a looming military disaster would have been intense and perhaps irresistible. The emerging nuclear taboo would have been broken and a major barrier to the further use of nuclear weapons greatly undermined.

2 · THE CHANGING TECHNICAL ENVIRONMENT

HURDLING THE TECHNICAL BARRIERS

Building the Bomb

The first technical barrier confronting a country that wanted to acquire nuclear weapons during the early decades of the nuclear age was mastery of basic nuclear knowledge and theory. That theoretical knowledge is increasingly accessible today. In fact, virtually all of the countries that might "go nuclear" in the years ahead—Iraq, Pakistan, South Africa, South Korea, and Taiwan, to name a few—already have the necessary theoretical knowledge.

These countries' scientists know, for example, that there are two isotopes or forms of uranium—U–235 and U–238—and that U–235 can be split (fissioned) when struck by a neutron, thereby releasing large quantities of energy and two or more additional neutrons. These neutrons, in turn, can split other U–235 isotopes, be absorbed by U–238 isotopes, or escape into the environment. Scientists also know the amount of fissionable material—or critical mass—needed for a controlled nuclear fission chain reaction as well as the conditions under which that material will react to produce the explosive chain reaction of an atomic bomb. And as the 1980s pass, more countries will cease being nuclear neophytes.[1]

Access to sufficient quantities of nuclear explosive material—that is, plutonium or highly enriched uranium—has been a second, more significant, barrier to countries that wanted to produce an atomic, or fission, bomb. Plutonium, which was used in the Alamogordo bomb, is the artificial by-product of the controlled fissioning of uranium. It is created by the capture of a neutron by the U–238 isotope and that isotope's subsequent radioactive decay. The Manhattan Project built a special production reactor to manufacture plutonium by "burning" uranium in a controlled nuclear chain reaction. Because plutonium is a by-product of nuclear power-generated electricity (which also involves a controlled nuclear chain reaction), it also can be diverted from the spent fuel rods of nuclear power reactors. Similarly, plutonium-bearing spent fuel rods can be diverted from a small nuclear research reactor. However, whether created in a production, power, or research reactor, plutonium must be separated chemically from the residual uranium and fission by-products and then processed into metallic form before it can be readily used in a bomb.

Of the two isotopes of natural uranium—U–235 and U–238—only U–235 can be used to make a fission bomb, and it is significantly less abundant. Uranium can be enriched, however, to increase the relative percentage of U–235 from less than 1 percent in the original ore found in nature to more than 90 percent. The resultant highly enriched uranium constitutes nuclear explosive material of the kind used in the Hiroshima bomb.

The number of countries capable of acquiring plutonium— either by illegal diversion from civilian nuclear energy activities or by building production facilities—is increasing steadily. By the mid-1980s, the nuclear research or energy programs of Argentina, Brazil, India, Israel, Pakistan, South Korea, Taiwan, and Yugoslavia all will have produced militarily significant quantities of plutonium.[2] Acquiring the needed facilities for separating that plutonium from the diverted spent reactor fuel is unlikely to be a major obstacle.

By the late 1970s, Argentina, Brazil, India, Israel, and Taiwan already had acquired either so-called hot-cells permitting the separation and safe handling of radioactive plutonium in the laboratory in a sealed compartment or even small reprocessing plants. Despite the Carter Administration's efforts to slow the spread of reprocessing technology, multilateral cooperation has been halfhearted, and countries still are able to purchase the necessary technology—as exemplified by Italy's decision in 1980 to go ahead with the sale of a large hot-cell to Iraq.[3] Moreover, many developing countries already are or will soon be able to design, engineer, and construct a crude but effective plutonium reprocessing facility within one to three years, drawing on the technical information available in the open literature.

A somewhat controversial 1977 study by the Oak Ridge National Laboratory concluded that a country with only a moderate technological base—defined as including commercial distilling, oil refining, and other chemical processing industries from which parts could be taken; a basic machine shop and metallurgical competence; and light construction equipment—could clandestinely build a "quick and dirty" spent fuel reprocessing plant in four to six months.[4] Other analyses question whether the plant could be kept secret and cite one to three years as a more realistic time period. But they agree with the basic contention that "Any state with some experience in building and operating complex chemical processes (oil refineries, for example) would have little difficulty in building a first pilot reprocessing plant to accept fuel from a commercial reactor."[5]

It is widely acknowledged as well that, by the mid-1980s, some or even many countries—including Brazil, Iraq, Pakistan, South Korea, and Yugoslavia—are likely to be capable of building at least a small natural uranium, graphite-moderated, air-cooled plutonium production reactor within three to four years.*[6] Comparable in design to the United States' Brookhaven Laboratory Graphite Research Reactor (the design and operating details of which are well known), such a reactor could produce enough fissionable material for several bombs per year. By the mid-1980s, the more technically advanced countries—such as Argentina, India, Israel, Spain, and Tai-

---

*A possible though not insurmountable problem for some countries in attempting to build either this or a larger production reactor could turn out to be access to uranium ore or other raw materials or components.

wan—probably will be capable of designing and building, in two to four years, a relatively large plutonium production reactor that could provide material sufficient for approximately twenty bombs per year. Moreover, there are other countries—such as Japan and West Germany—already capable of building such a large production reactor. And for all of these countries, separating the plutonium from the spent fuel is not likely to be a problem.

By contrast, only the more technically advanced countries, at least in the immediate future, are likely to be able to build a uranium enrichment plant to separate U–235 from U–238 and produce highly enriched weapons-grade uranium. South Africa, for instance, has built a small enrichment facility and is planning to expand it. But in the absence of significant outside assistance, such as Pakistan surreptitiously acquired for its clandestine centrifuge enrichment program, the technical and engineering barriers will be too great for most other countries.*

By the turn of the decade, gaining access to nuclear explosive material—and in large quantities—will be even less of a barrier. And though plutonium will remain the preferred route to the bomb, more countries will be capable of building an enrichment plant.[7]

Still a third barrier to those countries that wanted to join the nuclear club in the first decades of the nuclear age was designing and then fabricating the bomb. Design entails, for example, specification of the amount of nuclear explosive material required for a fission weapon, the means to reduce the required amount by increasing its density, and the mechanism for the rapid assembly of fissionable material sufficient to create a supercritical mass and, in consequence, an explosive chain reaction. In fabricating a plutonium bomb, particular care must be given to the design of the high-explosive lenses that create the

*France's sale to Iraq of a research reactor fueled with highly enriched uranium would have provided Iraq with direct access to a small amount of that material if it chose to violate its legal obligations to France. On this sale, see "France Plans to Sell Iraq Weapons-Grade Uranium," *Washington Post,* February 28, 1980; *Nucleonics Week,* March 6, 1980, p. 7.

supercritical mass by simultaneously compressing inward at all points a subcritical sphere of plutonium in a process called "implosion." Of equal importance is sufficient engineering, machining, and materials-handling skill to implement that design.

Design and fabrication of an atomic or fission weapon, too, are likely to pose few significant challenges to a growing number of countries. A great deal of formerly classified information concerning the principles and design of fission weapons now is available in the open literature.[8] In addition, the necessary technical skills in materials handling, precision machining, and metallurgy, among others, are also widely dispersed and will become more so. If fissionable material can be obtained, virtually all of the countries just mentioned—even those that are in the semideveloped category—in all probability will be readily able to design and fabricate a nuclear weapon in the 1980s.

Moreover, the first-generation nuclear weapons of most of the next countries that "go nuclear" may be considerably more sophisticated than the first American atomic bomb. Measuring 60 inches in diameter and weighing 10,000 pounds, the suitably labeled "Fat Man" plutonium bomb dropped on Nagasaki had a yield of approximately 14 kilotons (kt) or thousands of tons of TNT. With few exceptions, those countries that acquire nuclear weapons in the near future should be able to design and produce first generation bombs of 1,000 to 2,000 pounds, measuring between 20 and 35 inches in diameter, with yields of 20 to 200kt. With testing, these countries—ranging from Argentina to Taiwan—are likely to be capable of eventually producing even more efficient, more compact, and well-packaged fission devices.[9]

Delivering the Bomb

The final barrier to those countries that wanted to join the ranks of the nuclear powers during the first decades of the nuclear age was acquiring the capability to "deliver" nuclear weapons to chosen targets. This obstacle too is not as formidable as it used to be.

In the arsenals of Iraq and Israel, India and Pakistan, South Korea and Taiwan, as well as many, if not most, other countries that might acquire nuclear weapons in the decades ahead, are high-performance aircraft sold to them by the United States, the Soviet Union, or other countries. Any of these aircraft could carry a new first-generation fission bomb between 700 and 1,000 miles, a range sufficient to reach nearly all of the most likely regional targets of these countries and, in some cases, even beyond. During the 1980s, sales of even more advanced Soviet, American, French, and British aircraft will augment delivery capabilities. While such aircraft are not sold with the wiring and arming mechanisms necessary for dropping nuclear weapons, it will not be difficult for the recipient countries to make the required modifications.

As well, some of these countries will soon be producing their own nuclear-capable aircraft. Israel's recent development of the *Kfir* fighter and the growth of nascent aircraft industries in Argentina, Brazil, and India point toward an era of indigenously—or, where key parts are supplied from abroad, semi-indigenously—produced nuclear-capable aircraft.[10]

Some future nuclear weapons states also are likely to produce fairly compact, moderate-weight first- or second-generation nuclear warheads that could be carried by short-range surface-to-surface missiles (SSMs) rather than by high-performance aircraft. The developed countries are and will be the primary sources for such missiles. But Israel's production (with some help from France) of the *Jericho* SSM, recent South Korean missile tests, and research being conducted in Taiwan all suggest that, in the 1980s, a number of countries likely to acquire nuclear weapons may be capable of fabricating their own short-range missiles.[11]

For the more technologically advanced countries that will become capable of building even lower weight and more compact fission weapons—weighing on the order of 600 pounds—tactical surface-to-surface missiles, surface-to-air missiles reconfigured for surface-to-surface use, cruise missiles, and naval attack missiles and torpedoes* all could serve as delivery vehicles.

---

*Many countries that might be the targets of small power nuclear forces have coastal cities vulnerable to such a seaborne strike. The list of cities includes Buenos Aires, Rio de Janeiro, Alexandria, Tel Aviv, Bombay, Calcutta, Karachi, Algiers, Odessa, Vladivostok, New York, Los Angeles, and Boston.

The United States and the Soviet Union, as well as Israel and France, have sold missiles in this category to some countries that may acquire nuclear weapons. Capabilities for semi-indigenous production of less sophisticated variants on systems sold by the developed countries are also spreading.[12]

The development of sophisticated longer range missiles by the more technologically advanced countries is a further alternative. Both Japan, which already has a space rocket research program,[13] and West Germany, which has the building blocks for a missile program, are capable of developing intercontinental ballistic missiles (ICBMs), sea-launched ballistic missiles (SLBMs), or sophisticated cruise missiles.* Similarly, India, with its active space program (exemplified by its launching of a space satellite), should at least be capable of building intermediate-range ballistic missiles.[14]

Even large, unwieldy weapons of several thousands of pounds could be delivered by military transport aircraft or restructured commercial airlines,[15] both of which are possessed by virtually all sovereign countries today. These jerry-rigged systems could prove sufficient against local opponents with limited air defenses and could even be used on a suicide mission against one of the great powers, assuming successful subterfuge in penetrating its air space.†

As a last resort, a country, or even a subnational group, could attempt to smuggle a nuclear weapon into another country. The prospects for successful covert entry would depend heavily on the extent of border surveillance and customs checks. But in many developing regions there is widespread smuggling of goods across borders, while ton-loads of drugs

---

*Though such missiles are unlikely to match the systems by then available to the superpowers, even possession of 1975 state-of-the-art missiles could pose a formidable threat, unless sufficient countermeasures were taken by the superpowers.

†With its more developed air-defense network, the Soviet Union may be less vulnerable to this type of attack. On Soviet air defenses, see John M. Collins, *American and Soviet Military Trends since the Cuban Missile Crisis* (Washington, D.C.: The Center for Strategic and International Studies, Georgetown University, 1978), pp. 136–137.

are periodically smuggled into the United States. After successful placement, the nuclear weapon could be detonated by radio
command or by a preset timing mechanism.

Residual Technical Constraints

Technical barriers, although steadily becoming easier for many
countries to overcome, can still prevent some countries from
joining the nuclear club. Libya, for example, has been blocked
from producing nuclear weapons because it lacks access to di-
vertible fissile material and cannot produce that material itself.
And despite repeated attempts since the early 1970s, Libya has
been unable to circumvent these technical barriers by purchasing a bomb.[16] In the absence of outside assistance, technical
limitations could easily continue to doom Colonel Qaddafi's
quest for the bomb. Syria, Saudi Arabia, and Nigeria, to name
a few others, are similarly restrained.

Some countries capable of acquiring nuclear weapons, moreover, will find crossing the technical hurdles difficult, costly,
and time-consuming. Pakistan's efforts to acquire nuclear explosive material by building a uranium enrichment facility
have been under way for nearly eight years, with no immediate
end in sight.[17] Because of its limited manpower pool and technical base, Iraq's pursuit of the bomb was seriously slowed
when its chief nuclear scientist was murdered in Paris in June
1980 and when a year later Israeli aircraft attacked and destroyed Iraq's 40-megawatt nuclear research reactor.[18] Other
unexpected setbacks could range from periodic breakdowns of
reprocessing plants to accidents in the handling of nuclear
materials.

Residual technical constraints also may limit the number of
nuclear weapons produced by some new nuclear weapons
states. Similarly, the nuclear forces of many countries just
"going nuclear" may suffer from serious technical failings,
whether the lack of adequate measures to prevent the accidental detonation of a nuclear weapon, poor command and control, or high vulnerability to surprise attack. Technical difficulties also may prevent most of these countries from developing

more complex and sophisticated fusion (thermonuclear) weapons.*

Nevertheless, while technical constraints can still pose all these problems, they are not, in most instances, the stopping blocks that they used to be. And with the continuing global process of industrialization, scientific and technological development, and economic growth, overcoming the hurdles to the bomb will become even less of a challenge to an ever-growing number of countries. There is no way to stop this longer term change of the technical environment.

ATTEMPTS TO SLOW THE EROSION

There was great concern among U.S. policymakers in the mid-1970s that this longer term erosion was about to accelerate. In the wake of the 1973 oil price increases imposed by the Organization of Petroleum Exporting Countries (OPEC), many countries announced plans and took steps toward developing civilian nuclear energy. Pressure was building within the nuclear industries of the United States, Western Europe, and Japan to begin the commercial separation and use of plutonium

---

*Although thermonuclear weapons can be designed to have very low explosive power, their development also permits an order of magnitude jump in potential destructiveness from destruction measured in kilotons with fission bombs to destruction measured in megatons. On the average, fusion weapons also are more highly efficient, lighter weight, and more easily deliverable. In addition, the availability of "clean" thermonuclear weapons—designed for battlefield use, with less fallout—opens up greater flexibility of employment. However, projections of different countries' capabilities to build thermonuclear weapons are uncertain. It took France nearly seven years to test its first fusion bomb after development of its first fission bomb; China covered the same ground in less than three years. The preceding more cautious assessment of how long it will take countries in the next decades to develop fusion weapons draws on discussions with my late colleague at Hudson Institute, Donald G. Brennan, as well as with individuals at the United States nuclear weapons laboratories. Also see William Van Cleave, "Nuclear Technology and Weapons," in *Nuclear Proliferation Phase II*, ed. Robert M. Lawrence and Joel Larus (Lawrence: University Press of Kansas, 1974), pp. 54–55.

as a fuel for nuclear power reactors. France's decision in 1975 to sell Pakistan a large spent fuel reprocessing plant and West Germany's decision that same year to sell Brazil a complete civilian nuclear fuel cycle—technology for enriching uranium, nuclear power reactors, facilities for fabricating fresh fuel for those reactors, and a pilot-scale reprocessing facility[19]—were feared to presage intense commercial competition for sales of nuclear power reactors abroad. In particular, many experts worried that the major nuclear suppliers would use sales of "sensitive" plutonium reprocessing or enrichment facilities as sweeteners to induce countries to purchase their nuclear reactors. The sale of sensitive facilities would not only considerably reduce the difficulty of rapidly acquiring large quantities of nuclear explosive material, it would also erode the system of international controls governing the peaceful uses of nuclear energy. With national control of sensitive facilities, there would be too little warning after discovery of illegal diversion of plutonium for global diplomatic efforts to head off a nuclear test. Therefore, the central focus of U.S. nonproliferation policy under both Presidents Ford and Carter was to head off accelerated diffusion of such sensitive technologies while taking other steps as well to slow the longer term erosion of technical constraints.

Considerable efforts were made by both Administrations to convince France and West Germany to reverse their agreements to sell plutonium reprocessing plants to Pakistan and Brazil. In 1978, France refused to deliver the final components of the spent fuel reprocessing plant promised to Pakistan unless Pakistan agreed—which it did not—to a modification that would have rendered the plant of little use for a bomb program.[20] But despite American diplomatic arm-twisting throughout 1977 and into 1978, West Germany refused to go back on its agreement with Brazil.[21]

Between 1975 and early 1977, the United States also sought the adherence of the Soviet Union, France, West Germany, Canada, the United Kingdom, and Japan to tighten controls over the export of civilian nuclear technology. After considerable debate, formal agreement was reached in September 1977 on what came to be known as the London Nuclear Suppliers Group Guidelines, which were a limited but important step away from the danger of unprincipled and unbridled competition for nuclear exports.[22]

The Suppliers agreed, for example, that one condition for export of nuclear technology was the recipients' acceptance of safeguards, such as international inspection of nuclear facilities to ensure their use for only peaceful purposes. Failing to agree to a ban on sales of sensitive facilities such as the United States proposed, the Suppliers at least agreed to "exercise restraint"—as each country defines it—in exporting reprocessing or enrichment technology or facilities. The Suppliers also agreed to require a pledge from the recipient country, as a condition of export, not to use the transferred material, facility, or technology to make a nuclear explosive device. Moreover, any additional plants built within a specified time period that used principles or technology comparable to those of the original transfer were also required to come under safeguards. Further, as a condition of that initial transfer, recipient countries were required to agree to limits on their retransfer of such facilities as well as on the transfers by them of any indigenously produced facilities derived from the original transfers. The Suppliers also agreed that, in the event of suspected misuse of newly acquired civilian nuclear fuel cycle facilities or materials, they would consult about how to respond.

In April 1977, President Carter announced a series of major nuclear nonproliferation initiatives designed to create a new international consensus about the uses of nuclear energy. To set an example, he stated that the United States was deferring indefinitely the commercial reprocessing and use of plutonium and would attempt to find a more "proliferation-resistant" breeder technology. Incentives, including assurances of access to nuclear fuel, were offered countries to forgo acquisition of sensitive enrichment and reprocessing facilities. An international reassessment of civilian nuclear fuel cycle activities was also proposed.[23]

Spurred by Carter's call, many nations did come together under the rubric of the International Nuclear Fuel Cycle Eval-

uation (INFCE) between 1977 and 1980 to discuss how civilian
nuclear power might best be used while controlling the risk of
proliferation. Working groups were created and technical papers and opinions were exchanged on the availability of nuclear fuel, spent fuel reprocessing, breeders and other advanced nuclear power reactors, nuclear waste management,
uranium enrichment processes and requirements, and use of
nuclear power in the twenty-first century. Although the technical report issued by INFCE has been subject to many interpretations, it does, in part, support the Carter Administration's
argument against early and widespread commercial use of
plutonium.[24]

Paralleling these Administration efforts, the U.S. Congress
passed the Nuclear Non-Proliferation Act of 1978, specifying
conditions that had to be met before a country became eligible
for U.S. nuclear assistance or supplies. Among other requirements, recipients had to pledge not to make a nuclear explosive device, not to reprocess U.S.-supplied fuel without prior
approval, and to place all of their peaceful nuclear activities
under International Atomic Energy Agency (IAEA) safeguards
(so-called full scope safeguards). The legislation also named
specific conditions that would result in automatic termination
of American nuclear exports, such as detonation of a nuclear
explosive device, abrogation of IAEA safeguards, or contributing to the manufacture or acquisition of nuclear weapons by a
nonnuclear weapons state.[25]

AN ACCELERATED EROSION?

Backsliding by the Suppliers

These and other international efforts to check the erosion of
technical barriers could slacken during the 1980s. Renewed
commercial pressures for nuclear exports have prevented
agreement to extend the Nuclear Suppliers Guidelines and
may greatly weaken them. In early 1980, for example, West
Germany and Switzerland, afraid, respectively, of losing sales
of a heavy water power reactor and a pilot-scale plant for producing heavy water to Argentina, refused to make acceptance

of full-scope safeguards a condition of the sales. Moreover, by having arranged for the sale of the heavy water plant by a Swiss company with strong links to the West German firm of Kraftwerk Union, West Germany apparently circumvented a prior understanding with Canada, its main competitor for the reactor sale, that it would not sell such a plant without requiring full-scope safeguards. Both the West German and Swiss governments also rejected the U.S. contention that the heavy water facility was a sensitive facility, and thereby covered by the pledge to exercise restraint in the sale of such facilities in accordance with the Nuclear Suppliers Guidelines.[26]

Italy's readiness in 1980 to sell a hot-cell to Iraq as part of a broad effort to increase its commercial relations with that country also augurs a heightened danger of lessened restraint in sales of nuclear technology in the coming decade. The recent Soviet decision to help build up Libya's civilian nuclear infrastructure is equally disturbing. And while the new French government of President François Mitterand may seek to adopt a more restrictive nuclear exports policy, the same mixture of commercial security and foreign policy interests that led to France's 1977 sale to Iraq of a research reactor to be fueled by highly enriched uranium could erode that resolve.

Moreover, because of shifting priorities of the Reagan Administration, U.S. readiness to take the lead in pushing for exports restraint may decline. Costly controversies with key European allies are to be avoided. The result may be even greater weakening of suppliers' restraint.

Further, if the acquisition of nuclear weapons by additional countries in the early to mid-1980s leads—as it may, even if falsely—to the belief that runaway proliferation has become inevitable, these pressures are likely to intensify. The major nuclear suppliers will undoubtedly question why they should exercise restraint in selling "sensitive" reprocessing and enrichment facilities, particularly when these facilities will serve as effective sweeteners to win contracts for commercial nuclear power exports in an ever more competitive market. And a possible reluctance by the Reagan Administration to respond strongly to those initial proliferation outcroppings, say a Pakistani test that violated safeguards, would reinforce even more those pressures for nuclear laissez-faire.

Nuclear "Gray" and "Black" Marketing

Aside from increased backsliding by the major nuclear suppliers, the growth of nuclear "gray" and "black" marketing in the coming decade also threatens to make it still less difficult to hurdle the technical barriers to the bomb. The nuclear weapons activities of several countries already have been advanced in that fashion.

Pakistan's successful efforts to circumvent the network of nuclear exports controls set up by the major suppliers exemplify one type of nuclear gray marketing. From the mid-1970s to the early 1980s, Pakistan purchased components and equipment ranging from electrical frequency inverters to extremely tough high-quality steel for a clandestine centrifuge uranium enrichment facility.[27] Because these materials are used for both nuclear and nonnuclear activities, Pakistan was able to cloak its real intentions for some time. When the only use of a needed component was in an enrichment program, Pakistan relied on front companies, purchasing agents, and other subterfuges to conceal the ultimate destination of the equipment. Similar techniques have been used to obtain components in Pakistan's attempt to complete the spent fuel reprocessing plant bought from France and to build a smaller reprocessing facility on its own.[28]

Other countries may succeed in following Pakistan's example. Efforts to close this loophole are likely to be only partly successful because of the difficulties in coordinating international nuclear exports controls, the reluctance of some nuclear suppliers to adhere to these controls at the expense of business gains, the inherent ambiguity of some components and equipment, and the possibilities for concealing the purchaser's identity.

Another type of gray marketing is government-to-government assistance in the development of nuclear weapons between countries that are not party to the Treaty on the Nonproliferation of Nuclear Weapons. Cooperation may range from transfer of raw materials or components—perhaps even nuclear explosive material—to help in designing, engineering, building, and/or operating facilities for producing and processing plutonium or weapons-grade uranium, to joint research and development of missile delivery systems. Such cooperation may be motivated by a desire to solidify alliance ties, political pressures, or perhaps the need to acquire oil or another raw material. Joint ventures in nuclear weapons production are even possible.* For example, a country with sufficient scientific and engineering skills to develop a nuclear weapons program but lacking financial resources and/or access to the high-grade graphite needed in the manufacture of a plutonium production reactor might team up with a country that has ready financial resources and petroleum from which high-grade graphite can be made.

While there is no hard evidence of government-to-government nuclear weapons cooperation, there are sufficient hints, evidence of concern among officials, unconfirmed reports, and purported leaks to warrant taking this possibility seriously. Much recent speculation has focused on financial support by an Arab country—Saudi Arabia, Iraq, and Libya have been mentioned most frequently—for Pakistan's efforts to acquire nuclear explosive material.[29] The possibility of an Israeli–South African connection, entailing an exchange of nuclear weapons design information from Israel for uranium, or enrichment technology from South Africa,[30] or, possibly, Israel's access to a nuclear weapons test site in or off South Africa,[31] also has

---

*Problems with such joint activities should not be overlooked. Among the administrative and distribution problems of such ventures would be deciding where to locate the plant, assuring that once it was completed the host country would not simply be able to seize it, and dividing the output of the joint plant. Another source of difficulties, if the problems of the 1940–45 Anglo-American nuclear cooperation are a guide, would be controversy over who was making the greater contribution to the joint program and fears on the part of the more advanced partner that it was not being adequately compensated for its nuclear assistance. See Richard G. Hewlett and Oscar E. Anderson, Jr., *The New World, 1939–1946*, vol. 1: *A History of the United States Atomic Energy Commission* (University Park, Pa.: The Pennsylvania State University Press, 1962), pp. 255–258.

been widely discussed. Further, an Israel–South Africa–Taiwan consortium to produce cruise missiles is rumored to exist.[32]

The increasing availability of gray market "nuclear mercenaries" equally threatens to accelerate the breakdown of technical barriers. Ranging from out-of-work civilian nuclear engineers and technicians to former nuclear weapons laboratory employees, they may include physicists, metallurgists, chemists, and other needed technical specialists. Their motivation may be financial pressure, dissatisfaction with their current employment, or ideological affinity to the purchaser. Even if they lack nuclear weapons program experience, nuclear mercenaries may be able to provide theoretical knowledge and hands-on experience with metallurgy; reactor design, engineering, and operation; nuclear materials handling; spent fuel reprocessing; and similar civilian nuclear activities of relevance to a bomb program. Their past experience is likely to be valuable as well in organizing activities and in reducing the likelihood of costly problems and delays. And some nuclear mercenaries may even bring a knowledge of nuclear weapons design.

Nor is this mere speculation. Ronald Richter, the Austrian emigré scientist who tried to develop an atomic bomb for Argentina in the 1950s, was a forerunner of such nuclear mercenaries. More recently, it has been reported that some members of Israel's nuclear establishment have worked in the nuclear weapons programs of France and the United States.[33] In 1979, one of many reportedly dissatisfied Dutch nuclear scientists, engineers, and technicians is said to have signed on as a principal technical consultant for the reprocessing plant in Brazil.[34] Researchers at the Indian Bhabha Atomic Research Center, increasingly frustrated by bureaucratic shifts in support for their nuclear program, decreased funding, and overall lack of direction,[35] are prime candidates to become nuclear mercenaries. There also are reports of interest among Swedish nuclear specialists in possible openings in South Africa.[36]

Nuclear "black" marketing—the transfer or sale of diverted or stolen nuclear or even fissionable material—also threatens to speed up the erosion of technical constraints.* The theft and sale of nuclear material by a nuclear facility employee, criminal groups, subnational groups, or even the manager of a nuclear facility might be motivated by an ideological commitment to the recipient's cause, financial and personal gain, or a mixture of both.[37]

According to reports either leaked to the press in 1978 or made public under the Freedom of Information Act, the U.S. Central Intelligence Agency and the Defense Intelligence Agency concluded that, in the mid-1960s, enough highly enriched uranium to make ten fission bombs was diverted to Israel from an Apollo, Pennsylvania, fuel fabrication plant.[38] Even Joseph Hendrie, then chairman of the Nuclear Regulatory Commission, acknowledged in testimony to the House Subcommittee on Energy, in June 1979, that such a diversion might have taken place.[39] To circumvent safeguards, Israel also is thought to have "hijacked" a shipment of natural uranium in 1968, using false papers and bills of lading to conceal the shipment's destination.[40]

Additional gray and black market nuclear transactions are likely to take place during the 1980s and could become quite frequent if countervailing pressures weaken. Some low- to medium-technology countries would thus be able to initiate nuclear weapons programs earlier than otherwise would be possible, make more rapid progress toward their nuclear weapons objectives, and deploy nuclear forces of greater sophistication and magnitude. Iraq, in particular, may now rely heavily on the gray market to make up for the setback it suffered when its research reactor was destroyed in June 1981 by Israel. For still other countries, such as Libya, Saudi Arabia, and Nigeria, gray or black market nuclear transactions probably will be es-

---

*With the buying and selling of the services of nuclear mercenaries, the legal "color" of these proliferation-related activities clearly begins to shade from gray to black. Though possibly legal in the recipient country, selling of such personal services may be illegal in the mercenary's home country. For instance, U.S. law prohibits American citizens from providing such assistance to other countries without authorizations from the Secretary of Energy.

sential to their realization of any nuclear weapons aspirations in the next decade.

Third-Tier Nuclear Suppliers

The steps taken in the mid- to late-1970s to tighten controls over nuclear exports focused on the activities of the major nuclear suppliers. For example, both "first-tier" nuclear suppliers, capable of providing all nuclear fuel cycle services, technologies, and facilities, and "second-tier" suppliers, capable of exporting some, but not all, of those services, technologies, and facilities, signed the Nuclear Suppliers Guidelines.* But a "third tier" of nuclear suppliers has begun to emerge—comprised primarily of advanced developing countries, selling more technically modest components and facilities while providing training, expertise, and nuclear planning aid. Its growth also could accelerate the erosion of technical constraints on acquiring nuclear weapons.

The more recent activities of this third tier of nuclear suppliers have run the gamut from the training of foreign scientists at India's civilian nuclear centers[41] to Argentina's sale of a 10-megawatt (MWe) research reactor to Peru.[42] Other activities include Argentina's agreements to transfer nuclear technology and know-how to Uruguay, Paraguay, Colombia, Chile, Ecuador, and Bolivia;[43] Spain's agreement to sell a research reactor to Ecuador;[44] Brazil's agreement to transfer nuclear technology and training to Iraq[45] and its plans for further nuclear ties in the Middle East, South America, and perhaps even Africa;[46] plans for cooperation between the civilian nuclear industries of Brazil and Argentina;[47] and Pakistan's offer to provide training in the uses of nuclear energy for peaceful purposes to Malaysia and other developing countries.[48]

As the 1980s progress, the technical sophistication of exports by third-tier nuclear suppliers is likely to increase. Brazil as

*The London Nuclear Suppliers Group includes Belgium, Canada, Czechoslovakia, France, Democratic Republic of Germany, Federal Republic of Germany, Italy, Japan, the Netherlands, Poland, Sweden, Switzerland, the United Kingdom, the United States, and the Soviet Union.

well as Argentina, for example, may export small research reactors, while Argentina may begin to export larger research reactors and perhaps also heavy water and reprocessing technology to countries with only developing nuclear programs. And
as the years pass, still more countries—such as South Korea,
Taiwan, South Africa, and Yugoslavia—are likely to qualify as
third-tier nuclear suppliers.

Existing nonproliferation regulations and guidelines only
partially cover the activities of third-tier nuclear suppliers.
Many of these countries are not parties to the NPT and are not
bound by that Treaty's requirement that "sensitive" and certain
other types of nuclear technology be sold only if covered by
safeguards. The Nuclear Suppliers Guidelines include provisions that deal with the retransfer of technology or materials
originally transferred from one of the major suppliers, but
there are likely to be disputes about whether or not a third-tier
supplier's exports were derived from that original transfer. Besides, the Guidelines do not cover indigenously developed facilities and technology or those dependent on assistance from
another third-tier supplier or on gray market transactions.

For the most part, this new group of suppliers has accepted
the principle of nuclear exports restraint, but this situation
may not hold true much longer.[49] Many of these countries suffer from domestic economic difficulties compounded by the financial strain of an adverse balance of payments, which provide a strong incentive for using nuclear exports as sweeteners
for broader trade packages made up of manufactured or basic
industrial goods, construction services, raw materials and commodities, and, in many cases, conventional arms. Brazil's 1980
deal to transfer nuclear training and technology to Iraq, which
is part of just such a package and was clearly motivated by the
need to bring Brazil's exports and imports into closer balance,
is a case in point.[50] Some third-tier suppliers' commitments to
continued restraint probably will be weakened also by their dependence on oil from countries seeking nuclear technology.
While Libya's attempts in 1978 and 1979 to use its oil to gain
access to sensitive nuclear technology from India appear to
have failed, there are indications that Brazil may eventually

yield to Iraq's requests for such assistance.[51] Moreover, some countries, such as Argentina, regard nuclear cooperation with other third-tier suppliers as a way to reduce their dependence on the major suppliers and to circumvent the restrictions on civilian nuclear energy activities imposed by them.[52]

Still another source of pressure on restrained behavior is the perceived need by some third-tier suppliers for political support. An increasingly isolated South Africa, for example, may seek to trade access to unsafeguarded enriched uranium or even enrichment technology for diplomatic support. Or Pakistan might conclude that to preserve and extend its ties with other Moslem countries it has no choice but to help them at least to acquire sensitive facilities, if not nuclear explosive material. Should U.S. bilateral relations with any one of these third-tier nuclear suppliers worsen—as is occurring with India because of strengthened U.S. ties to Pakistan—this could engender a greater readiness to respond to such pressures.

Lessened restraint on the part of third-tier suppliers could weaken technical constraints well out of proportion to these countries' levels of nuclear sophistication. A growing alternative source of sensitive technology, components, and experience could make for more rapid possession of a larger number of nuclear weapons by some countries. Moreover, faced with the export of sensitive nuclear technology by third-tier nuclear suppliers, one or more of the major nuclear suppliers may no longer be willing to abstain from using such transfers as sweeteners for their commercial nuclear sales and other exports.

A Wasting Asset

Nuclear weapons are an old technology. Nearly four decades since the dawn of the atomic age, a growing number of countries can match the feat of the U.S. Manhattan Project. And backsliding by the major suppliers, increased nuclear gray and black marketing, and the emergence of third-tier nuclear suppliers all could accelerate that erosion of technical constraints in the years ahead. Technical barriers clearly are a "wasting asset" for nonproliferation.

3 • HOW MANY COUNTRIES WILL GET THE BOMB?

SLOW AND LIMITED PROLIFERATION CONTINUES

The continued spread of nuclear weapons, covertly or overtly, appears unavoidable. In South Asia, the Middle East, and Southern Africa, several countries already are actively seeking a nuclear weapons capability. Nonetheless, contrary to more pessimistic predictions, the pace of that proliferation may be slow and its scope limited, much as in the first decades.

India and Pakistan: "If Them, Us"

Soon after his assumption of power in January 1972, Prime Minister Zulfikar Ali Bhutto set in motion efforts to develop a Pakistani bomb.[1] These efforts were accelerated considerably in the wake of India's 1974 nuclear test. In 1975, Pakistan ordered a French plutonium reprocessing plant. Under pressure from the United States, however, the French government decided in 1978 not to provide the components needed to complete the plant. Otherwise, this plant most likely already would have been used by Pakistan to separate plutonium from a civilian nuclear power reactor for a nuclear weapons program. A statement smuggled from Bhutto's jail cell during his 1978 trial by the military government of General Mohammed Zia ul Haq that had overthrown him provides virtual confirmation:

> We know that Israel and South Africa have full nuclear capability. The Christian, Jewish and Hindu civilisations have this capability. The Communist powers also possess it. Only the Islamic civilisation was without it, but that position was about to change.[2]

44

The collapse of the deal with France stalled Pakistan's pursuit of nuclear weapons but did not derail it. Steps were taken to complete the French reprocessing facility and to build another smaller facility as well.[3]

Moreover, in the mid-1970s, unnoticed at the time by Western intelligence agencies, Pakistan began building, at Kahuta, a clandestine centrifuge uranium enrichment plant capable of producing weapons-grade uranium.[4] Pakistan relied heavily on gray market purchases of components and materials for the facility. This program also benefitted considerably from technical information, reports, and blueprints supplied by Dr. Abdul Qadar Khan, a Pakistani metallurgist who, because of a security breach, was permitted access to confidential documents at the URENCO enrichment plant in Almelo, the Netherlands, in 1975.[5] By mid-1978, when Pakistan's efforts were uncovered, the centrifuge enrichment program was too far along to stop. U.S. intelligence specialists reportedly estimate that, barring an unexpected technical failure, the Kahuta facility could provide weapons-grade uranium sufficient for a nuclear explosive test by the early 1980s.[6]

In part, Pakistan's activities have been motivated by the fear that, should India deploy a full-fledged nuclear force, it could use that force to coerce Pakistan into accepting its political demands in a crisis.[7] Equally important has been Pakistan's belief that nuclear weapons could help compensate for the weakness of its conventional military forces[8]—a belief bolstered by the view that possession of only a few nuclear weapons, or even the capability to develop them quickly, might lead to a more stable relationship, based on mutual nuclear deterrence, with India.[9]

National pride and a quest for greater international status have reinforced Pakistan's security incentives for acquisition of nuclear weapons. Detonation of a nuclear explosive device would demonstrate its technical and scientific prowess, both to the Pakistanis themselves and to skeptical outsiders—expecially to the Indians. It also would set Pakistan apart. In the words of one Pakistani official, "The Iranians have oil, Indians have Carter as well as the Device—it's only fair that Pakistan should have at least the bomb."[10] And as the first Moslem country to acquire nuclear weapons, Pakistan would achieve heightened prestige among those Moslem nations whose political and economic support it is actively seeking.

Domestic political considerations increasingly have played a role in Pakistan's continuing efforts to "go nuclear." The shaky government of General Zia has ignored outside pressures in the past few years to discontinue its nuclear weapons activities, in part because that would have caused popular, elite, and even military protests and disaffection. Moreover, in the absence of other policy successes, acquisition of a nuclear explosive capability, if not the detonation of a device, may be the only way for Zia to prop up his domestic position.[11]

How far Pakistan will go along the nuclear road is somewhat uncertain. At the very least, once technically ready, Pakistan is likely to begin stockpiling plutonium or weapons-grade enriched uranium while continuing research on the design and fabrication of nuclear weapons. Perhaps equally likely, Pakistan could covertly fabricate and stockpile nuclear weapons components, stopping just short of final assembly of nuclear weapons. But particularly in light of reports that it is preparing a test site,[12] Pakistan may be unwilling to stop without openly testing at least one nuclear explosive device—possibly claimed to be a "peaceful nuclear explosive" (PNE). Or it may even take the next step of producing nuclear weapons and seeking to deploy a full-blown nuclear force.

The covert acquisition of untested or unassembled nuclear weapons or other activities short of full-fledged deployment might satisfy most of Pakistan's incentives for "going nuclear." Rumored possession of untested nuclear weapons, or, even more so, a single detonation, may be thought sufficient to establish a deterrent relationship with India, assuage domestic opinion, buttress Zia's legitimacy, and establish Pakistan as a leader of the Moslem world. Nevertheless, Pakistan's military leaders may decide to push ahead not only to a nuclear test but also to full-fledged military deployment of nuclear weapons. Perhaps surprisingly, concern in Pakistan about India's response to either action is not very evident; nor is there much fear of the impact of a costly nuclear weapons program on Pakistan's already weak conventional defenses. To a considerable extent, Pakistan's decision will depend on its leaders' perceptions of both the benefits of Pakistan's new political, economic, and security relationship with the United States and the risk of its disruption if Pakistan "goes nuclear."

India's May 1974 nuclear test was not followed by a decision to produce nuclear weapons, and by 1977 the Indian government of Prime Minister Morarji Desai had shelved India's nuclear weapons program.[13] The initial response of both the Desai government and that of Indira Gandhi (who returned to power in January 1980) to Pakistan's efforts to acquire nuclear explosive material combined watchful waiting with warnings of readiness to carry out "without hesitation" further nuclear explosions if national interest so demanded.[14] The limited threat posed by Pakistan's activities, skepticism about Pakistan's technical capabilities, and the potential international political and economic costs of resuming India's activities all contributed to this restraint.

By the early spring of 1981, however, pressures for taking India's nuclear weapons program off the shelf seemed to be mounting. Warnings of a readiness to resume testing became more frequent, while there were reports of renewed activity at India's Pokharan nuclear test site.[15] Even so, India's response may continue to be restrained once Pakistan begins to operate the Kahuta facility or completes the French or the smaller semi-indigenous reprocessing facility. Should Pakistan detonate a nuclear explosive device in 1982 or 1983, however, or even begin to stockpile untested bombs, it is likely to trigger a full-scale resumption of India's awakening nuclear weapons program.[16] In part, India would act out of concern that failure to match Pakistan's nuclear weapons would place it at a political and military disadvantage in a crisis or conflict with Pakistan. Unwillingness to accept a position of less than clear-cut military superiority vis-à-vis Pakistan—which would clash with India's long-standing belief in its legitimate regional preeminence—also would play an important role.[17] Further, Pakistan's actions would have reduced the political costs of India's resuming its nuclear weapons program.

Whatever Pakistan does, there will be strong pressures in the

1980s for India to resume its nuclear weapons activities. India's international and domestic situation today closely resembles that of May 1974. Once again Mrs. Gandhi may conclude that a nuclear test is a way of "shaking her fist" at the United States for its favoring Pakistan over India and of distracting attention at home from economic difficulties and domestic unrest. Besides, the drive for political independence and global recognition and influence that originally contributed to India's decision to detonate a nuclear device retains a powerful appeal to the elite.[18] As well, growth of China's nuclear capabilities may be thought by India to require fresh measures of its own, and the continuing modernization of China's conventional forces, with Western assistance, is likely to heighten India's interest in buttressing its deterrence of conventional incursions through the deployment of nuclear weapons.[19] Finally, by the mid- to late-1980s, all the components needed for a militarily significant Indian nuclear weapons program will be in place—ready access to significant quantities of indigenous nonsafeguarded nuclear explosive material, medium-range missiles developed in India's space program, organizational skills, trained manpower, and electronics.[20] Taken together with the other incentives, scientific momentum is likely to tip the balance in favor of a full-fledged program.[21]

If India decides to resume its nuclear weapons program, it will place nearly irresistible pressure on Pakistan to produce and deploy nuclear weapons. And disincentives to doing so are likely to be outweighed by the perceived risks of inaction.

Overt Proliferation in the Middle East

More likely than not, Israel already has covertly produced a limited number of assembled or nearly assembled atomic bombs. As early as 1974, the Central Intelligence Agency concluded that Israel had the bomb.[22] Among the circumstantial evidence that subsequently persuaded virtually all U.S. government officials and outside analysts to agree with that conclusion are Israel's reported clandestine acquisition of weapons-grade uranium, its refusal to allow inspection of the Dimona research reactor, Israeli intelligence leaks about the assembling of atomic bombs during the first days of the October 1973 war when Israel's survival was threatened, its tough-minded readiness to take all steps necessary for its defense, its building of missiles designed to accommodate nuclear warheads, Israeli air force exercises of tactics required to drop atomic bombs, and possible access to data from France's nuclear weapons tests.[23] The Israeli government's periodic statement that "Israel will not be the first country to introduce nuclear weapons into the Middle East"[24] certainly does not preclude possession of unassembled components that could be turned into bombs in a matter of a day or two, if not hours. Indeed, in June 1981, former Israeli Defense and Foreign Minister Moshe Dayan acknowledged that Israel "ha[s] the capacity" to build the bomb "in a short time."[25] And there are compelling incentives for covert acquisition of nuclear weapons, including the desire to hedge against unexpected military reverses or shifts in the Arab-Israeli balance of power, a belief in the value of a last-resort nuclear threat either to deter direct Soviet military intervention or to induce U.S. military assistance lest nuclear weapons be used in the Middle East, and the possible deterrent effect of nuclear weapons possession on Arab leaders.[26]

Pressures that may lead to Israel's reassessment of its decision to stop short of overt testing, the production of increasingly efficient and sophisticated weapons, and deployment of a nuclear force are building.[27] Continuing U.S. sales of high-performance aircraft and other advanced military equipment to Egypt and Saudi Arabia[28] are already seen in Jerusalem as lessening Israel's present conventional military superiority and are likely to heighten security-related incentives to demonstrate an overt nuclear capability. The possible collapse of Egyptian-Israeli peace talks and the restoration of a hostile Arab front also would increase security-related pressures, as would still further reduced confidence in the United States as a guarantor against large-scale direct Soviet intervention. Equally important, the deepening economic disruption stemming, in part, from high conventional defense spending is stimulating debate about reliance on a "more bang for the buck" nuclear deterrent posture.[29] But the greatest pressure is the fact that by the late 1980s, if not sooner, one or more Arab countries are likely to acquire nuclear weapons.

On June 7, 1981, Israel attacked and destroyed Iraq's nuclear research reactor, Osirak, which was about to begin operation. This raid disrupted what both Israeli and U.S. officials increasingly believed was an attempt by Iraq—using its leverage as an oil supplier—to acquire the necessary components and materials for a nuclear weapons program under the guise of building up peaceful nuclear activities.[30] And when Iraqi President Saddam Hussein in his first public comment after Israel's attack called on all other countries to help the Arabs get nuclear weapons, that seemed to confirm these suspicions.[31]

These suspicions began soon after Iraq signed an agreement with France in 1977 to purchase the Osirak nuclear research reactor—to be fueled with highly enriched uranium and to be supplied with several extra reloads of fuel. If illegally diverted, in violation of international safeguards, the enriched uranium would have sufficed to produce at least several nuclear weapons. Before these reactors could be shipped to Iraq, their cores were sabotaged mysteriously in April 1979—allegedly by Israel. France offered Iraq replacement cores that operated with non-weapons-grade uranium, but Iraq refused to accept the substitution. Although France then acquiesced and supplied the original-style cores, it did reduce the number of fuel reloads initially supplied.[32]

In 1980, Italy agreed to provide a large hot-cell to Iraq, despite protests from the United States that the hot-cell could be used to separate plutonium from spent fuel as part of a nuclear weapons program.[33] There was also discussion about the purchase of an Italian heavy water research reactor, capable of producing plutonium for a bomb program.[34] The same year, Brazil—another country dependent on Iraqi oil—signed an agreement to pursue joint uranium exploration with Iraq and to assist Iraq in operating its research reactor, ensuring necessary safety, and conducting basic nuclear research. There has been speculation, fueled by "misstatements" by Brazil's ambassador to Iraq, that Brazil might be willing to transfer sensitive reprocessing information, hands-on experience, and technol-

ogy to Iraq,[35] all of which would be useful for a plutonium production program and would have contributed to any Iraqi attempt to acquire nuclear explosive material by misusing the
French research reactor. At about the same time, Iraq purchased nonsafeguarded natural uranium from Portugal, Niger,
and possibly Brazil.[36] If Osirak had become operable, that uranium could have been used to produce enough plutonium for
about one Hiroshima-type bomb per year, assuming the Iraqis
illegally reconfigured Osirak by placing a blanket of uranium
around the core—a project not beyond Iraq's expanding engineering and industrial capabilities, particularly if Iraq used its
ample financial resources to purchase components on the gray
market.

Israel's raid, however, has delayed and derailed but not
ended Iraq's apparent quest for the bomb. Even if the Mitter-
and government does not replace Osirak, there are other
routes to nuclear explosive material. Perhaps Iraq's most likely
course is to build a secret, Manhattan-Project-type plutonium
production reactor, making use of its nuclear ties to Brazil and
buying gray market components, materials, or expertise as
needed.

There are indications that Libya too is still seeking nuclear
weapons despite its technical backwardness. Using its leverage
as an oil supplier, Libya tried unsuccessfully throughout 1979
to pressure India into providing sensitive technology and
know-how for a nuclear weapons program.[37] There have been
persistent reports (of uncertain credibility) that Libya is funding Pakistan's enrichment program,[38] in return for which it
may expect to receive anything from Pakistani-built centrifuges
to the transfer of nuclear explosive material. Against this back-
ground, Libya's efforts—with Soviet assistance—to build up an
indigenous civilian nuclear base also must be a cause of
concern.[39]

It is likely that both Iraq and Libya believe that acquisition
of at least a nuclear weapons option is a legitimate and necessary response to what Hussein and Qaddafi are convinced is
Israel's covert possession of nuclear weapons. Possession of
even untested nuclear weapons might deter nuclear blackmail

by Israel, while a nuclear test may be seen as a useful diplomatic tool to shatter the increasingly unacceptable Middle East status quo. According to Iraq's or Libya's scenarios, faced with the threat of nuclear destruction in the Middle East and losses of access to oil, the superpowers would intervene to avoid the outbreak of conflict and impose an overall peace settlement on Israel. Besides, acquisition of nuclear weapons in all probability is thought a useful support for Iraq's emerging claim to a preeminent role in Persian Gulf and Middle East affairs,[40] while also serving Colonel Muammar Qaddafi's quest for regional and international influence. And by its preventive attack on the Osirak reactor, Israel has reinforced the incentives of Iraq and Libya, if not other Arab countries, to acquire nuclear weapons. Not only has the attack affronted national pride but it has demonstrated the potential political utility of the bomb— for Israel clearly has shown that it will feel threatened by Arab possession of nuclear weapons.

Because of their importance as oil exporters, both Iraq and Libya probably are doubtful that there will be much adverse Western reaction to their "going nuclear"—even if they violate international safeguards. They also may assume that any Soviet response would be tempered by a reluctance to sacrifice its other foreign policy ambitions on the altar of nonproliferation. But Israel's possible reaction is a more compelling disincentive, particularly after the June 1981 attack that Prime Minister Begin has threatened to repeat as needed.[41] And even though some questions were raised in Israel about the timing of the raid,[42] there should be little doubt that Israel will try to make good its threat to derail a revived Iraqi, a Libyan, or any other Arab nuclear weapons program before it is completed. But such a preventive strike might fail—whether because of heightened Iraqi secrecy, dispersal and hardening of Iraqi facilities, or improved defenses.

If Iraq or Libya successfully acquire a limited number of untested bombs by the mid-1980s, Israel also may stop short of unveiling its nuclear weapons capability—in part to avoid alienating the United States and for fear of providing the Soviets an excuse for more active involvement in the region. But it is much more likely—consonant with Israel's warnings that it will not stand aside to such nuclear weapons development—that Israel will take steps to maintain its clear-cut nuclear superiority through the development, production, and military deployment of thermonuclear weapons. Successful Iraqi or Libyan testing and deployment would almost certainly result in overt Israeli deployment of nuclear weapons and heighten pressures elsewhere within the Persian Gulf and Middle East for acquisition of nuclear weapons—not least of all in Egypt, Syria, Jordan, Saudi Arabia, and Iran. But except for Egypt, these countries probably will find it extremely difficult to cross the technical barriers to the bomb, barring those barriers' accelerated erosion and, in the case of Iran, restored political and economic stability.

South Africa: How Close to the Bomb?

In early August 1977, photographs taken first by Soviet and then by U.S. intelligence satellites revealed an apparent nuclear weapons test site in the Kalahari Desert. These photographs spurred widespread fears of an imminent South African nuclear weapons test that would use highly enriched weapons-grade uranium produced in South Africa's newly operational pilot-scale enrichment facility.[43] A combined American, British, French, West German, and Soviet diplomatic offensive was mounted to head off that test.[44] And on August 23, President Carter stated:

> South Africa has informed us that they do not have and do not intend to develop nuclear explosive devices for any purpose, either peaceful or as a weapon; that the Kalahari test site, which has been in question, is not designed for use to test nuclear explosives; and that no nuclear explosive test will be taken in South Africa now or in the future.[45]

But in an interview with a U.S. television network one month later, South African Prime Minister Vorster denied having provided such assurances. He claimed that he had only "repeated a statement which I have made very often that, as far as South Africa is concerned, we are only interested in peaceful development of nuclear facilities."[46]

Just over two years later, concern over South Africa's possible nuclear weapons activities was greatly heightened. On September 22, 1979, a U.S. VELA satellite, put into orbit to monitor compliance with the 1963 ban on the atmospheric testing of nuclear weapons, sighted two bright pulses of light in the South Atlantic,[47] the sequence and timing of which closely resembled the signature of a nuclear weapons detonation.[48] But despite an intense search by the United States and several other countries, no other unambiguous geophysical evidence of a nuclear test—such as radioactive debris, passage of a shock wave, electromagnetic or light radiation, individual eyewitness reports, or radiation in the upper atmosphere—was uncovered.[49] Nevertheless, it is possible that a nuclear test occurred without such confirming evidence. An expert panel set up by Carter to resolve the matter described a chain of highly improbable natural circumstances that could explain the double flash, but also admitted that its explanation was not fully credible.[50] The CIA and the Defense Intelligence Agency, scientists at the U.S. nuclear weapons laboratories, and the Navy's Oceanographic Laboratory disputed the panel's findings.[51] To this day, the initial presumption that a nuclear explosive test occurred remains neither proved nor disproved.

Rumored possession of possibly tested nuclear weapons is a useful bargaining tool for South Africa. Western opposition to the demands of African countries for the diplomatic and economic isolation of South Africa derives, in part, from concern that one consequence probably would be South Africa's overt testing and deployment of nuclear weapons. More generally, such rumors signal both the neighboring black African countries and their Soviet and Cuban supporters that the risks and costs of military action against South Africa may be quite high.[52]

South Africa's actual deployment of a small nuclear force capable of threatening neighboring countries would visibly demonstrate its readiness to resist with all means their demands for black rule in South Africa and quite possibly would weaken those countries' resolve to support the military overthrow of South Africa's white regime. It also could buttress deterrence of military intervention in Southern Africa by the Soviet Union and Cuba. Even though any South African use of nuclear weapons against Cuban or Soviet forces undoubtedly would be answered by devastating retaliation, the Soviets still might fear that events would get out of hand and that if the destruction of their regime appeared imminent, South Africa's leaders might take desperate or irrational actions beyond what the Soviets would be willing to risk.[53] Should South Africa be able to mount a credible threat to one or more Soviet cities—for example, a crude cruise missile launched from a submarine or a short-range missile on a surface craft—the deterrent effect would be considerably enhanced. Overt deployment of nuclear weapons also might be thought to provide added insurance against the South African government's fears of a large-scale, externally supported conventional attack.[54] Not least of all, overt testing of nuclear weapons also would buttress the morale of the white population in South Africa and demoralize the black opposition by demonstrating to both that the government intends to stand behind the apartheid regime at all costs.[55]

While actual overt testing and deployment of nuclear weapons is consistent with South Africa's emerging "Fortress Southern Africa" defense strategy—the tightening of the laager to meet heightened external and internal threats[56]—it also entails significant risks, among them increased international isolation, a damaging oil or general economic embargo, augmented legitimacy for Soviet or Cuban military intervention in the region, and the complete breakdown of South Africa's longstanding effort to forge closer security ties with the West.[57] Thus, South Africa may continue to eschew an overt nuclear weapons program; conversely, if that government perceives that its security is eroding and finds itself under siege anyway, in the early to mid-1980s South Africa most likely will proceed to build a full-fledged nuclear force.

Nigeria's defense minister responded in 1980 to South Africa's rumored nuclear weapons activities with the statement that "as long as the protagonists of apartheid have access to nuclear capability, Nigeria should, of necessity, endeavor to acquire it at any price."[58] But Nigeria's efforts—as well as those of any other black African states—will be hindered by its limited technical, scientific, and industrial infrastructure. So unless the erosion of technical constraints accelerates, whether due to the availability of extensive gray and black market assistance or a breakdown of suppliers' restraint, South Africa will probably keep its regional nuclear monopoly.

STILL MORE ACTIVITY ALONG THE NUCLEAR ROAD?

While a continuation of the first decades' proliferation pattern, with four or five more countries acquiring nuclear weapons covertly or overtly in the next decade or so, is the most likely outcome, it is far from the only possible one. If the erosion of technical constraints accelerates, Nigeria, Saudi Arabia, and Syria, to name a few countries with possible motivations but hampered by technical backwardness, all could "go nuclear" by the end of the decade. Moreover, should certain political and economic conditions change, still other countries probably would join the ranks of the nuclear club. Slow and limited proliferation would not be the outer boundary but rather the initial baseline of future proliferation.

Rekindling Dormant Programs in Northeast Asia

In 1975, Taiwan's President Chiang Ching-Kuo acknowledged that studies and preparations for a nuclear weapons program were begun by his country in the late 1950s and that by 1974 Taiwan had acquired the technical capability to manufacture nuclear weapons.[59] Taiwan's nuclear weapons program had begun as a response to China's nuclear capability, but increasingly became a hedge against an erosion of U.S. security ties.[60] Under economic and political pressure from the United States, and fearful of provoking that very erosion, Taiwan terminated its nuclear weapons program in 1977. It agreed to close and seal an experimental plutonium reprocessing facility that reportedly had secretly reprocessed small quantities of spent fuel, vowed not to engage in activities related to reprocessing, and accepted restrictions on the use of a 40-megawatt nuclear research reactor of the type that had produced the nuclear explosive material for India's 1974 nuclear test.[61] Despite the U.S. decision in December 1978 to recognize mainland China and to terminate the 1954 Mutual Defense Treaty, there has been no public evidence of a resumption of Taiwan's nuclear weapons program. Apparently, the residual U.S. security guarantee, reinforced by concern about jeopardizing Taiwan's economic and political ties with the United States, has so far sufficed to prevent that step.

South Korea's initial efforts to acquire a nuclear weapons capability were apparently triggered in 1970, when the Nixon Administration announced that one division of U.S. ground forces would be withdrawn from South Korea. The Korean Weapons Exploitation Committee was formed, and it voted, sometime in 1970 or 1971, to go ahead with the development of nuclear weapons, sending missions abroad to gather information and components. These preparations were both a bargaining lever to stave off future troop withdrawals—lest South Korea "go nuclear"*—and a hedge against those withdrawals. In late 1975, the United States, with Canadian support, successfully brought strong pressure to bear on South Korea to terminate its nuclear weapons activities.[62] But had the Carter Administration in July 1979 not reversed its 1977 decision to withdraw a second U.S. division from South Korea,[63] security pressures to reestablish South Korea's nuclear weapons program would have been intense.

These underpinnings of Taiwan's and South Korea's nuclear weapons restraint could erode during the 1980s. China already has objected to continued U.S. sales of conventional weapons

---

*Thus high South Korean officials, including President Park, periodically have explicitly linked the American security guarantee and nuclear weapons abstinence. "Faute de la protection américaine, nous fabriquerions nos propres armes nucléaires déclare le Président Park," *Le Monde* (Paris), June 14, 1975; "Official Hints South Korea Might Build Atom Bomb," *New York Times,* July 1, 1977; Young-sun Ha, "Nuclearization of Small States and World Order: The Case of Korea," *Asian Survey* 18, no. 11 (November 1978), p. 1142.

to Taiwan[64]—sales important for Taiwan's defense capability and a symbol of the residual U.S.–Taiwan security link.[65] Political shifts within China or the United States could lead to heightened Chinese demands for more rapid resolution of "the Taiwan issue." Even barring those changes, China might make still lessened U.S. support for Taiwan a precondition to improved Sino-American relations and possible defense cooperation against the Soviet Union. Further, the Reagan Administration's emphasis on greatly strengthening the capability to project U.S. conventional forces into the Middle East and the Indian Ocean, combined with budgetary constraints and a continuing serious shortage of trained U.S. military personnel, may yet revive pressures for redeploying U.S. ground forces now stationed in South Korea. Continued political unrest and disturbances in South Korea also could lead to the loosening of U.S. security ties in an effort to avoid entanglement in a military conflict brought about by Korea's instability.

Any such reduction of the U.S. security presence in Northeast Asia will greatly heighten the prospects for resumed proliferation in that region.[66] To the leaders of Taiwan and South Korea, even a small number of nuclear weapons might again seem necessary to buttress their countries' bargaining positions in a future political confrontation, improve public and elite morale, avoid the emergence of a deterrence gap, and enhance their defenses in the event of invasion.

Even assuming a marked intensification of incentives to "go nuclear," Taiwan and South Korea will have difficulty deciding how and how openly to acquire nuclear weapons. Fear of preemption, of severing of American diplomatic and military support, and of loss of access to foreign peaceful nuclear technology would remain pressing disincentives to overt deployment of nuclear weapons. Consequently, both Taiwan and South Korea initially might go no further than covert acquisition of untested devices. Their decisions would depend heavily on whether that more limited step was thought sufficient to satisfy their respective political and security requirements.

Even if Taiwan and/or South Korea resume their nuclear weapons programs, those actions probably will not lead to an

immediate chain reaction in the region. Because of North Korea's more modest nuclear, industrial, and technical base, without outside assistance its efforts to match its neighbor's nuclear
weapons program would be quite drawn out. A still-valid 1977
estimate by the former U.S. Energy Research and Development Administration, for example, did not even include North
Korea among countries capable of acquiring nuclear weapons
within seven to ten years of a decision to do so.[67] Also, despite
some speculation to the contrary,[68] a South Korean decision to
"go nuclear" would probably be too weak a political or security
shock to overcome Japan's powerful disincentives to acquiring
nuclear weapons.

The Breakdown of Nuclear Restraint in Latin America

During the mid-1970s, many observers feared that the mutual
suspicion of Argentina and Brazil, their traditional rivalry, and
each country's quest for prestige would lead them to acquire
the bomb.[69] Brazil's purchase of sensitive reprocessing and enrichment technology from West Germany in 1975 had fueled
those fears and led to warnings from Argentina that, without
assurances from Brazil, it might have to pursue nuclear weapons.[70] Argentina's decision in 1978 to build a full-scale reprocessing facility seemed a step in that direction.[71]

In the past year or two, however, the suspicion between Argentina and Brazil has decreased markedly.[72] In part, this is
the consequence of the overall improvement of Argentine–
Brazilian relations that followed the 1979 inauguration of General João Figueiredo as president of Brazil.[73] The first stages of
civilian nuclear cooperation—from exchanges of information
and personnel to purchases of components—between Argentina and Brazil also helped lessen uncertainty about each
other's nuclear programs.[74] Besides, because of problems and
delays, Brazil's nuclear program seemed far less threatening.

Of equal importance, there is far less consideration in Brazil
today than there was several years ago of increasing Brazil's
world role and influence—an objective that many thought demanded nuclear weapons. Increasingly, Brazil is focusing on
serious domestic problems that run the gamut from a massive

balance-of-payments deficit (due to Brazil's heavy dependence on imported oil) to the sociopolitical instabilities that accompanied the Figueiredo Administration's relaxation of authoritarian rule.[75] At the same time, Argentina's lingering desires for increased global prestige and status seem to have been satisfied by its expanding nuclear diplomacy, including ties to many Latin American states and a broader claim to leadership among third-tier nuclear suppliers and others dissatisfied with the existing restrictions on access to civilian nuclear technology.

But neither Brazil nor Argentina has been willing to give up its nuclear weapons option, whether by ratifying the Treaty on the Nonproliferation of Nuclear Weapons, accepting full-scope safeguards, or pledging not to make nuclear explosives. And although Brazil, unlike Argentina, has formally ratified the Treaty of Tlatelolco, creating a nuclear weapons-free zone in Latin America, it is unwilling to agree to implement the treaty unless Argentina, Cuba, and the existing nuclear powers also adhere to the treaty's restrictions. Both countries have paid a political, technical, and economic price to avoid limiting their nuclear weapons options.

Domestic political considerations have played a role in these decisions. Neither country wanted to give in publicly to the United States on the nuclear issue. But a residual mutual uncertainty and rivalry between Argentina and Brazil has also been important.[76] Should their nuclear cooperation turn sour— say in a dispute over the extent of information being exchanged or the equitable sharing of new projects—it could exacerbate the lingering distrust.

Moreover, Brazil's goal of achieving global power and status has only been displaced, not discredited, by Brazil's forced preoccupation with domestic matters.[77] Many proponents and supporters of increasing Brazil's international standing are still close to the seats of power; if Brazil's domestic problems are brought under control by the late 1980s, their voices may again be influential. Conversely, a severe domestic breakdown could bring power to a populist-nationalist regime that would revive Brazil's foreign policy ambitions. In either case, the result may be a Brazilian nuclear weapons program, placing great pressure on Argentina to follow suit.

With its traditional volatility and unpredictability, Argentina itself could change course and take steps to acquire nuclear weapons in the next decade. Once other medium powers such as India, Iraq, or South Africa acquire the bomb, Argentina's deep-seated need for status and prestige might motivate it to do likewise.[78] Or, faced with growing domestic economic and political difficulties, the Argentine junta might adopt a more outward, nationalistic foreign policy—including building the bomb—to distract public attention from those problems.

But even if there is such a breakdown of nuclear restraint in Argentina and Brazil, runaway proliferation throughout the region still might be avoided. Technical constraints will greatly hinder matching efforts by smaller neighbors such as Uruguay, Peru, Colombia, and Cuba. More important, assuming that Argentina and Brazil stop short of stockpiling large numbers of nuclear weapons or at detonation of a so-called peaceful nuclear explosive, security- and status-related pressures in Chile, Venezuela, and Mexico could be more limited than expected. Besides, for some of these countries, concern about adverse American or Soviet reactions, loss of access to advanced nuclear technology, and possible problems in assuring control of nuclear weapons in politically unstable countries all may help tip the balance in favor of a more restrained response.

Yugoslavia

Although Yugoslavia took some initial steps toward acquiring nuclear weapons in the early 1960s, it signed and ratified the NPT in 1970. A controversial article that appeared in *Borba,* the Communist party newspaper, in 1975 seemed to reopen the issue. It argued that Yugoslavia should adopt all necessary means for its defense, including battlefield nuclear weapons.[79] Later disavowed by its author, this article apparently outran officially approved speculation about a Yugoslav bomb, but a second article, published in 1977, did not. Its author, Colonel-General Ivan Kukoc, a ranking Yugoslav Army general,

warned that Yugoslavia's nonnuclear weapons position could
change if conditions so warranted and stressed that continued
discrimination against Yugoslavia in the civilian nuclear field
would heighten pressures for a reassessment of nuclear policy.[80] According to some observers at the time, publication of
the Kukoc article reflected the growing interest of a segment
of the Yugoslav armed forces in reconsidering acquisition of
nuclear weapons to enhance deterrence of the Soviet Union.[81]

In the mid- to late-1980s, Yugoslav interest in nuclear weapons may revive still further. Even a few nuclear weapons—
either deliverable against a Soviet city or for use on the battlefield—could be thought a desirable buttress to deterring Soviet
use of force to bring Yugoslavia back into the bloc now that
President Tito is dead. Nuclear weapons also might be seen as
necessary to stiffen elite resolve and to make up for the loss of
Tito in Yugoslav-Soviet dealings. Besides, not only would acquisition of nuclear weapons by Yugoslavia increase the risks
to its neighbors of cooperating with Soviet military action, it
is also likely to make them less willing to pursue irredenta of
their own. Further, the acquisition of nuclear weapons could
be conceived as helpful to the pursuit of national unity and
sovereignty in times of heightened domestic political unrest in
Yugoslavia.

Until now, fear of the Soviet response has been an over-
whelming deterrent to Yugoslav acquisition of nuclear weapons. The Soviet Union has made it clear to the countries of
Eastern Europe that efforts to acquire nuclear weapons would
provoke dire consequences.[82] But in the not improbable event
of a Soviet succession crisis in the mid- to late-1980s, countries
on the periphery of Soviet influence and control could find
their freedom of action increased. Heightened preoccupation
of the Soviet leaders with domestic affairs and jockeying for
power could lower significantly the risks of a small covert nuclear weapons program in the eyes of the Yugoslav leadership.
Once Yugoslavia made public its covert production of even a
limited number of nuclear weapons, it might be hard for the
Soviets to reverse that fait accompli without running excessive
risks. Yugoslavia then could pursue a more substantial overt

program under the nuclear umbrella provided by those secretly acquired bombs.

Iran Catches Up

There was considerable concern about Iran's nuclear-weapons
ambitions in the 1970s. Not only had Iran begun an ambitious
nuclear power program and sought access to sensitive technologies, but also, in 1975, the late Shah of Iran stated that "if
ever a country [of this region] comes out and wants to acquire
atomic weapons, Iran must also possess atomic bombs."[83]

The fall of the Shah, however, considerably reduced the
danger of such a regional chain reaction. The Islamic regime
of the Ayatollah Ruhollah Khomeini does not share the Shah's
expansive definition of Iran's security interests or his pursuit
of regional and global preeminence—either of which would
have provided compelling arguments for pursuing atomic
weapons following decisions by India and Pakistan to "go nuclear." Further, even if Iraq, which poses a serious and continuing security threat to Iran, should overtly test and deploy nuclear weapons in the mid to late 1980s, Iran probably will not
have the capability to follow suit. Iran's nuclear research and
energy program has been dismantled; many of the trained
managers, technicians, and engineers needed for a nuclear
weapons program have fled the country; the economy is in a
shambles; and domestic political instability shows little sign of
abating and may intensify.

Nevertheless, though unlikely, Iran's growing domestic economic, administrative, and political decay could be checked in
the decade ahead. For example, as has happened before in
Iran, the conservative mullahs may overstep themselves and
trigger a backlash by Westernized technocrats and business-
men. Or a new military leader, with the support of a nationalist
civilian-military junta, may take power. There is little doubt
that such a new government would attempt to catch up with
Iraq's nuclear weapons activities, not only out of concern about
nuclear blackmail or attack but also in competition for regional
influence. This new government also might believe that military prudence required not falling too far behind India and

Pakistan. Depending on its political complexion, it even might renew the Shah's quest for global prestige. But while the return of effective government would make possible serious efforts to catch up, lingering damage done by the Khomeini order as well as residual technical constraints would make it unlikely that Iran could acquire the bomb before the decade ends.

RUNAWAY PROLIFERATION

The first decades' pattern of slow and limited proliferation might even break down completely in the years ahead, driven by renunciation of nuclear abstinence by Japan and West Germany. Though these countries' acquisition of nuclear weapons admittedly is much less likely than the preceding developments, equally improbable events have occurred throughout history, surprising observers and confounding confident soothsayers. A change of posture by Japan or West Germany most probably also would engender a belief that runaway proliferation no longer was avoidable and would fundamentally undercut support for nonproliferation policies globally.

Japan: The Typhoon Mentality

The prevailing consensus in Japan is that the costs of nuclear weapons acquisition outweigh the benefits.[84] The public's nuclear allergy, which dates from Hiroshima and Nagasaki, remains an important disincentive. Notwithstanding growing domestic support for increased conventional defenses,[85] there is only extremely limited political sentiment favoring nuclear weapons acquisition—nor is that likely to change in the foreseeable future. The economic costs of a large-scale, sophisticated nuclear weapons program—one that includes submarine-launched missiles that could attack cities in the western Soviet Union—are thought by Japanese analysts to be too high. And though difficult to believe, some of those analysts even question Japan's technical and organizational capability to carry out such a large-scale nuclear weapons program.

Widespread skepticism about the benefits for Japan's security or foreign influence of possessing nuclear weapons also feeds into the nonnuclear consensus. The Japanese are quite ready to rely on the U.S. security guarantee and are unwilling to contemplate seriously the withdrawal of the American nuclear umbrella. Besides, there is considerable doubt about the utility of nuclear weapons to meet Japan's specific security threats—whether, for example, a growing Soviet presence in the region, interference with the shipping of goods and raw materials to Japan, a dispute with China or Vietnam over offshore oil, or renewed conflict on the Korean peninsula. It is also said that should South Korea or Taiwan acquire nuclear weapons, this will not lead Japan—either for security reasons or for reasons of status—to follow suit.

Nonetheless, many of those Japanese observers who argue that a rational calculation of costs and benefits dictates continued abstention also admit that Japan might overreact to an external security shock by launching a nuclear weapons program. For, to paraphrase one Japanese defense analyst, the Japanese are a typhoon people, making few preparations before sudden shocks; but, when the shock (typhoon) comes, they are very surprised and then go ahead and do more than necessary.[86] And there are many possible shocks.

Although Japan does seem likely to adjust readily should South Korea or Taiwan develop nuclear weapons, the actual testing and deployment of nuclear weapons might prove far more unsettling to Japanese self-esteem and prestige than anticipated. Another security shock capable of changing Japan's nuclear weapons policy could be a complete severing of U.S.–South Korean security ties, perhaps in response to the emergence of South Korea as a nuclear weapons power. It is also possible, though less likely, that a combination of growing U.S. bitterness over the costly domestic impact of Japan's protectionist economic policies, increasing U.S. retrenchment in Asia (perhaps simply out of the need to redeploy forces elsewhere), and rising Japanese nationalism will result in the breakup of the American-Japanese security alliance in the mid-1980s. Equally unnerving to the Japanese would be an American defeat outside of Asia, calling into question the value of

U.S. guarantees. Nor is it possible to rule out an incident during the next decade involving Japan and the growing Soviet military forces in Asia, in which U.S. assistance either would be ineffective or simply not forthcoming. Should any of these security shocks cause Japan to develop nuclear weapons, it will certainly increase the pressure on other countries in the region—and, in some cases, beyond—to do the same.

West Germany

In all probability, West Germany will not choose to acquire nuclear weapons in the coming decades. One reason for this is fear of a hostile Soviet response—possibly military in nature, but certainly political and economic. Acquisition of nuclear weapons also would alienate West Germany's Western European neighbors and undermine more than thirty years' effort to earn readmittance to the European community.[87] Most important, as long as it believes its security is adequately assured by the NATO alliance and the U.S. nuclear guarantee, West Germany will not have a compelling incentive to shift course.

But there are certain possible shocks that might cause West Germany to rethink or question its reliance on the security framework provided by NATO. West German perceptions of increasing Soviet strategic superiority,[88] for example, could lead to more vigorous questioning of the alliance's effectiveness and of the American guarantee. A breakdown of current NATO efforts to implement a coordinated military and political response to the heightened threat posed by the continuing modernization of Soviet theater nuclear forces, and especially deployment of the SS-20 missile, also could greatly erode alliance cohesion.[89] Renewed pressures for the withdrawal of U.S. troops from Europe, fueled by clashes within NATO over fair sharing of the burden of defense, would have a comparable adverse impact. Further, a major American defeat in a confrontation or clash with the Soviets—perhaps in the Persian Gulf—could also shake West Germany's confidence. Conversely, fears that any Soviet-American conflict on the peripheries would spill back into Western Europe might raise doubts about the U.S. security connection. Or, a unilateral

American declaration that it would not use nuclear weapons first would greatly increase West German insecurity, given NATO's reliance on the threat of nuclear first use as well as West Germany's emphasis on deterring conflict by threatening nuclear escalation.

These or similar security shocks are a necessary, but possibly not sufficient, provocation for West Germany's renunciation of its nonnuclear posture in the coming decades. The threat of hostile Soviet and West European reactions still will be a powerful countervailing force. It is also possible that West Germany's leaders, faced with a more uncertain security environment, might attempt to strike a deal with the Soviet Union—withdrawing from NATO and maintaining military neutrality in return for Soviet political guarantees, closer ties to East Germany, and long-term economic arrangements. Also possible would be West German support for the transformation of France's nuclear force into an intra-European nuclear deterrent—although reliance on a French nuclear umbrella not only has practical drawbacks but is likely to be even more uncertain than reliance on an eroded U.S. guarantee.[90]

Should West Germany acquire nuclear weapons, at least some other European countries will be pressured to do the same, whether for reasons of prestige in the case of Italy, out of a sense of intensified insecurity for many Eastern European countries, or as a reluctant adjustment to the heightened role of nuclear weapons in Europe. Particularly if Japan acquired nuclear weapons at roughly the same time, the already weakened belief that runaway proliferation is avoidable would probably collapse altogether. And by acquiring nuclear weapons, those two countries would fundamentally undermine any remaining nonproliferation efforts. For faced with such an irretrievable failure, the major industrial countries are likely to ask why they should continue to bear the risks, burdens, and potential commercial sacrifices of standing behind the nonproliferation regime. As a consequence, those countries that had not "gone nuclear" before for fear of sanctions would have less reason to hold back and the position of opponents to the acquisition of nuclear weapons will be weakened. Still other

countries with nuclear weapons programs in the early stages
will undoubtedly accelerate their efforts. Finally, as the erosion
of technical constraints accelerates, efforts to enforce agreed-
on nuclear rules of the road among the suppliers also will
slacken.

Thus, the driving forces that could lead to runaway proliferation are clear. The erosion of American alliances, the
growth of an "if them, us" logic in situations where local rivals
fear each other's nuclear weapons intentions, the intensification of prestige and status incentives, the domestic political
utility of a nuclear weapons program, scientific and bureaucratic momentum, declining fear of adverse international reaction, reduced domestic opposition, and an acceleration of the
erosion of technical constraints—all would increase the pace
and scope of proliferation.

Nonetheless, there is no inexorable dynamic leading inevitably to the complete breakdown of the first decades' pattern of
slow and limited proliferation. Nor is there reason to conclude
that once two or three more countries, or even nine or ten, acquire some form of nuclear weapons capability there will be no
stopping point until virtually every sovereign nation possesses
the bomb. Potential firebreaks exist, both regionally and globally. The outcome will depend on somewhat uncontrollable political and economic developments abroad but even more on
the policies pursued by the United States and other countries.

4 • WHAT DIFFERENCE WILL IT MAKE?

THE BREAKDOWN OF NUCLEAR PEACE

A number of analysts and observers, noting that predictions at the dawn of the nuclear age of a nuclear apocalypse have proved exaggerated, argue that there is little reason to fear the consequences of the further spread of nuclear weapons.[1] Frequently at the core of such optimistic assessments is the belief that the very destructiveness of those weapons will both instill prudence in their new owners, making them less willing to use even minimal conventional force out of fear that conflict will escalate to use of nuclear weapons, and lead to stable deterrent relationships between previously hostile countries.[2] But such a fear of nuclear war was only one of the underpinnings of the first decades' nuclear peace. Other equally significant geopolitical and technical supports may be absent in the conflict-prone regions to which nuclear weapons are now likely to spread.

Higher Stakes, Shorter Distances, and Failures of Leadership

The stakes of the long-standing rivalries and conflicts in the Middle East, South Asia, the Persian Gulf, and the Korean peninsula—contrasted to those of the superpower confrontation—are very high. Territorial integrity, political independence, and, in some instances, even national survival itself frequently are at issue. Consistent with the magnitude of the stakes, leaders of countries in these regions have proved willing in the past to use military force against their regional opponents—whether, for example, in an attempt to push Israel into the sea or to unify Korea, to dismember Pakistan and create Bangladesh, to topple a rival leader in a neighboring country, or to seize new or regain lost territory. And because they perceive the stakes to be so high, some of these countries' leaders may be ready to risk nuclear confrontation, if not even to accept a surprisingly high level of nuclear damage, in pursuit of their objectives. Thus, it would be erroneous to assume that these new nuclear powers will necessarily subscribe to the Western "minimum deterrence" point of view—that the threat of one or two atomic bombs dropped in retaliation on an opponent's capital city will suffice to deter the use of force and prevent lesser clashes.[3]

Regardless of these leaders' intentions, flash points for conflict among these new nuclear powers abound. The festering civil war in Lebanon involving Christians, Palestinians, Syrians, and Israelis; border clashes between Libya and Egypt; a renewal of the Iraq-Iran war; a new incident between the Koreas; and unrest in Baluchistan or Kashmir are all potential tripwires. And the risk of unintended escalation will be considerable. For unlike the Soviet Union and the United States, many of the next countries that may "go nuclear" share common borders or are separated only by narrow or unstable buffer zones. Once under way, limited confrontations or low-level clashes could spill over quickly into vital national territory and threaten critical national interests, perhaps even survival. Further, again in contrast to the superpowers, little if any time may be available for learning to live with nuclear weapons before the first such nuclear confrontation occurs.

Moreover, the legacy of conflict among these countries, as well as their domestic political weaknesses, could make it all the more difficult to check such a slide to all-out war.[4] In a climate of deeply rooted hostility and distrust, the leaders of, say, Israel and Iraq may refuse to make concessions during a crisis, fearing that they will be considered weak and thus subjected to further demands. Or, fearful of being thrown out of power, a weak leader could be reluctant to take the first step to defuse a crisis until it was too late. For example, the weakened Pakistani government of Yahya Khan in 1971 could not bring itself to meet East Pakistani demands for greater autonomy, result-ing in the third Indo-Pakistani war and Pakistan's dismember-ment. And even leaders made considerably more prudent by the threat of nuclear war can miscalculate how far they can push in a crisis, just as Nasser did when his rhetorical postur-ing and decision to move troops into the Sinai helped trigger the 1967 Middle East war. Similarly, although offered neutral-ity by Israel, King Hussein decided to enter the same war—at the cost of Israeli occupation of the West Bank.*

Besides, judging from recent history, it is highly unlikely that coldly calculating, cautious, and fully rational leaders—at least in the Western sense—will always be in authority in the next countries to acquire nuclear weapons. In many of these countries, the volatility of domestic politics, the psychological and personal strains caused by vast economic and social changes, and the weakness of political institutions will result in the periodic seizure of power by extremist military cliques, messianic leaders, and fundamentalist religious and ideological movements. Leaders such as the Ayatollah Khomeini, Colonel Qaddafi, and Pol Pot, committed to the pursuit of transcenden-tal goals and societal redemption, exemplify what can be ex-pected. Motivated by an obligation to their higher mission and destiny, these leaders are less likely to weigh carefully the costs and gains of military action and are more likely to take high-risk gambles to serve their causes.

Technical Deficiencies of New Nuclear Forces

Considerable time, money, and scientific and engineering tal-ent have been spent by the United States since the 1950s to de-sign, develop, and implement accident-proofing for opera-tional nuclear weapons.† Efforts to insure that these weapons

---

*Both Shai Feldman and Kenneth Waltz, for example, not only assume that new nuclear powers will accept the logic of minimum deterrence, but they greatly underestimate the dangers of escalation.

†The initial American mode of accident-proofing employed for the atomic strike against Hiroshima was to assemble the final components of the bomb only after the aircraft was airbone. But in the 1950s, after the United States began to deploy large numbers of nuclear weapons in a nuclear deterrent force, more sophisticated measures integral to the warhead design itself also were developed.

could, for example, withstand the heat and impact of air crashes and be dropped accidentally without producing a nuclear explosion paid off in the 1950s and 1960s, when nearly two dozen American aircraft crashed while carrying nuclear weapons.[5]

Some of those countries that may soon acquire nuclear weapons, however, probably do not have the scientific and engineering manpower or the financial resources to design and fabricate reliable, advanced accident-proofing systems integral to the nuclear weapon itself. Their limited resources will be expended in their struggle simply to join the nuclear club. Fearful of being caught unprepared, however, these countries may be unwilling to use such a simple, if less sophisticated, accident-proofing measure as not fully assembling their nuclear weapons until they are needed. And even stockpiling disassembled weapons would not reduce the risk of an accident once those weapons have to be assembled. In fact, hasty assembly of crude weapons would augment the chances of just such an accidental detonation in the most dangerous political context—a continuing crisis or even a low-level conflict.*

Some of the next countries to "go nuclear," unable, say, to build invulnerable missiles, also will be forced to rely on a hairtrigger, launch-on-warning operating procedure in attempting to protect their nuclear force from surprise attack. This will heighten the risk of an accidental nuclear exchange triggered by mechanical failure or human error.† There will be virtually

---

*Some proponents of the benign consequences of nuclearization of regional conflict, such as Paul Jabber and Steven Rosen, assume that this problem of accident-proofing would be solved by the enlightened transfer of technical assistance from the superpowers. That begs the question. For political and technical reasons, the superpowers might decide not to provide such assistance. Nor is it clear that "assistance" that would require access to a new nuclear power's weapon design would be accepted by those countries themselves. Other optimistic assessments, including Shai Feldman's, simply neglect this problem.

†Both mechanical and human error, for example, occurred in the course of breaking in both the Distant Early Warning (DEW) Line and the American Ballistic Missile Early Warning System (BMEWS) in the 1950s. However, other technical and geopolitical characteristics did not reinforce that error as might happen for new nuclear powers. See Joel Larus, *Nuclear Weapons Safety and the Common Defense* (Columbus: Ohio State University Press, 1967), pp. 37–38.

no time to verify the initial warning of attack because of the short distances separating most of these new nuclear powers; but where the stakes are high, pressure to act on such a warning lest a surprise attack succeeds will be intense.

Even with reliance on launch-on-warning, it is highly probable that at least a few of these new nuclear forces will remain vulnerable to surprise attack, decreasing significantly an opponent's aversion to nuclear escalation. For example, in a confrontation between India and Pakistan in the late 1980s, in which Pakistan relies on aircraft for the delivery of a handful of nuclear weapons while India has available nuclear-armed missiles, an Indian nuclear first strike might virtually destroy Pakistan's nuclear force, thereby greatly reducing the threat of nuclear retaliation. Or, Israel probably would have a similar advantage in a confrontation with Iraq, Libya, or Egypt. For while these countries have unsophisticated surface-to-surface rockets, technical constraints will probably impede their development of small nuclear warheads for them. Thus, unless widely dispersed and concealed,* Iraqi, Libyan, or Egyptian bombers would be highly vulnerable to attack by Israeli nuclear-armed missiles.† In still other regional nuclear confrontations, say between Libya and Egypt or Iran and Iraq, both sides might be vulnerable to surprise attack.

Inadequate controls against the unauthorized use of nuclear

---

*Command and control problems might make widespread dispersal of nuclear-armed aircraft unattractive to an Iraqi or Egyptian government unsure of its own military; if pursued anyway, such dispersal might solve the first strike problem but exacerbate that of unauthorized access. And even if nuclear-armed aircraft were dispersed, Israel could use its nuclear weapons to blanket large areas with sufficient over-pressures from the nuclear blast to destroy above-ground missiles. Predictions of stable deterrence as the probable outcome of Middle East nuclearization skip over such difficulties.

†Steven Rosen, Paul Jabber, and Shai Feldman, for example, apparently rest their expectation of stable deterrence in a nuclearized Middle East on the prospect that at least one or two warheads would survive an Israeli surprise attack and that this minimum threat would suffice to deter an Israeli first strike. But that depends on the stakes involved as well as on perceptions of the risk that events would get out of hand. And, in a low-level Arab-Israeli conflict, striking first and accepting the risk of such retaliation might come to be viewed by Israel as its least undesirable course of action.

weapons by disaffected military men—or against the theft of those weapons by dissidents and terrorists—is likely to be a further technical weakness of some newly deployed nuclear forces even though measures to reduce that threat are in theory available. One such control measure accessible to countries with a certain amount of technical sophistication is placing an electronic permissive action link (PAL) on each nuclear weapon, thereby requiring a particular code to "unlock" and use that weapon. In less technically advanced countries, special civilian forces could guard disassembled nuclear weapons stored at some distance from their delivery vehicles. Still another possibility would be centralized storage of all nuclear warheads at one or two sites, again removed from delivery vehicles and guarded by elite military units. But in addition to a heightened vulnerability to surprise attack, the penalty for reliance on such less technically sophisticated measures is significantly decreased operational readiness and effectiveness, as it would take considerable time to remove the nuclear warheads from storage, ship them to forward bases, and "mate" them to their delivery vehicles.

In practice, some of the new nuclear powers that are unable to develop electronic control measures most likely will find the penalties of relying on less sophisticated measures unacceptable. More fearful of surprise attack, they may choose to accept the risks of unauthorized use inherent in the decentralized storage of nuclear warheads in close proximity to their delivery vehicles or in the advanced mating of warheads and delivery vehicles.* Consequently, for these countries—a group that could soon include Pakistan, Iraq, and Libya—maintaining control over nuclear weapons will depend heavily on the effectiveness of physical security procedures. A special "civilianized" military guard might complicate efforts by a few military men or a subnational group to seize one or more nuclear weapons but could be readily overwhelmed in a military coup d'etat. Be-

---

*Recent more optimistic evaluations of the consequences of proliferation, including those of Kenneth Waltz, Steven Rosen, Paul Jabber, and Shai Feldman, either neglect the command-and-control problem or simply assume that it will be solved.

sides, members of that guard might be bribed, coerced, or persuaded to cooperate.

The significance of these command-and-control deficiencies is magnified by the domestic political instability of most of the countries that may acquire nuclear weapons by the early 1990s. Nearly all—including Argentina, Brazil, Egypt, Iran, Iraq, Libya, Nigeria, Pakistan, and South Korea—are developing countries, while many either have experienced a successful or aborted military coup d'etat within the past decade or could in the future.[6] This lack of military subordination to higher authority would increase the importance of rigorous command-and-control measures. But for some of these countries' leaders a recognition of the need for such measures in all probability still will be outweighed by the perceived security need of an operationally ready—even if less than tightly controlled—nuclear force.

A Spiraling Threat to Peace

With the spread of nuclear weapons to conflict-prone regions, the chances that those weapons will be used again increase greatly. The heightened stakes and lessened room for maneuver in conflict-prone regions, the volatile leadership and political instability of many of the next nuclear powers, and the technical deficiencies of many new nuclear forces all threaten the first decades' nuclear peace.

Not least to be feared is nuclear war caused by accident or miscalculation. During an intense crisis or the first stages of a conventional military clash, for example, an accidental detonation of a nuclear weapon—even within the country of origin—or an accidental missile launch easily might be misinterpreted as the first shot of a surprise attack. Pressures to escalate in a last-ditch attempt to disarm the opponent before he completes that attack will be intense. Similarly, a technical malfunction of a radar warning system or a human error in interpreting an ambiguous warning might trigger a nuclear clash. Or fear that escalation to nuclear conflict no longer could be avoided might lead a country to get in the first blow, so as partly to disarm the opponent and to minimize damage.[7]

Unauthorized use of nuclear weapons by the military also is a possibility. For example, faced with imminent conventional military defeat and believing there is little left to lose anyway, a few members of Pakistan's military could launch a nuclear strike against India to damage that country as much as possible. Or a few hard-line, fanatic Iraqi, Libyan, or even Egyptian officers might use their countries' newly acquired nuclear weapons in an attempt to "solve" the Israeli problem once and for all. These officers' emotional commitment to a self-ordained higher mission would overwhelm any fear of the adverse personal or national consequences. Aside from the initial destruction, such unauthorized use could provoke a full-scale nuclear conflict between the hostile countries.

But the first use of nuclear weapons since Nagasaki may be a carefully calculated policy decision. The bomb might be used intentionally on the battlefield to defend against invasion. For example, faced with oncoming North Korean troops, a nuclear-armed South Korea would be under great pressure to use nuclear weapons as atomic demolition land mines to close critical invasion corridors running the thirty miles from the border to Seoul. Similar military logic could lead to Israeli use of enhanced radiation weapons—so-called neutron bombs—in the next Arab-Israeli war.

A calculated disarming nuclear surprise attack to seize the military advantage also is possible in these high-stakes, escalation-prone regional conflicts, particularly when one side has a decided strategic advantage. For example, in the 1980s, internal political instability in Pakistan and simmering unrest in Kashmir could erupt into a conventional military clash between India and Pakistan. A nuclear-armed India then would be under intense pressure to attack the more rudimentary Pakistani nuclear force to prevent its use—whether by accident or intention—against India. In a nuclear Middle East, as well, fear of events getting out of hand would fuel arguments in favor of an Israeli first strike once a conflict had begun.

Aside from the increased threat of actual use of nuclear weapons, the nuclearization of conflict-prone regions may have other costly or dangerous consequences. Given the stakes, some new nuclear powers will think seriously about a preventive strike with conventional weapons to preserve their regional nuclear monopoly. Israel already has taken such military action against Iraq's nuclear weapons program and has stated its readiness to take further action as needed. And notwithstanding the limited Iraqi reaction to Israel's preventive strike—in large part due to Iraq's being tied down in its war with Iran— it might not be possible to prevent escalation after similar or larger future attacks. (Though less likely, a sufficiently desperate country might even use nuclear weapons in such a preventive strike. There is evidence that in the late 1960s the Soviet Union seriously considered a preemptive nuclear strike against the nascent nuclear force of the People's Republic of China.)[8]

Possession of nuclear weapons also may be used as an instrument of blackmail or coercion. A country with a nuclear edge may implicitly or explicitly threaten the use of nuclear weapons to enforce its demands in regional crises or low-level confrontations. Just as U.S. strategic superiority contributed to the Soviet Union's decision to back down in the 1962 Cuban Missile Crisis,[9] so might possession of nuclear weapons by Iraq, Israel, India, or South Korea affect the resolution of crises with weaker opponents.

In addition, tensions among the countries of newly nuclearized regions are likely to be exacerbated. Pakistan's nuclear weapons activities, for example, already have heightened India's suspicion and have slowed efforts to improve relations between the two countries.[10] Pakistani testing and deployment of nuclear weapons would further worsen relations between India and Pakistan, not least because such activity would affront India's claim to regional preeminence. Should India step up its nuclear weapons activities in response and achieve clear-cut nuclear superiority, Pakistan's fears of Indian nuclear blackmail would be increased as well. Even the anticipation of a country's "going nuclear" can have adverse political effects. For example, Iraq's efforts to acquire nuclear weapons have heightened Israel's siege mentality and stimulated efforts by Syria, Saudi Arabia, and even Kuwait at least to master basic nuclear theory and know-how.

The greater the scope, the quicker the pace, and the higher the level of proliferation, the more severe will be the threat of nuclear conflict. As more countries acquire the bomb, the number of situations in which a political miscalculation, leadership failure, geographical propinquity, or technical mishap could lead to a nuclear clash will increase. As the pace of proliferation accelerates, the time available for countries to adjust to living with nuclear weapons will grow shorter. As countries move to the more advanced levels of proliferation—from untested bombs to full-fledged military deployment, there is more chance that some of these new nuclear forces will be technically deficient. Further, nuclear weapons will cease to be isolated symbols and will become an integral part of international relations within these volatile regions.

There will be occasional exceptions where the spread of nuclear weapons does not have as adverse an impact as feared. For example, if Taiwan's acquisition of nuclear weapons does not provoke an immediate preventive strike by China, it might go far toward eliminating a military solution to the conflict between them. Or a Yugoslav nuclear force capable of surviving a Soviet disarming attack and destroying several Soviet cities in retaliation could offer needed deterrence against a Soviet military incursion. And even though acquisition of nuclear weapons by Argentina and Brazil probably would heighten mutual suspicion and political tensions between them, the risk of actual nuclear conflict is likely to be less than in other regions because of the more modest stakes of their traditional rivalry.

The initial outcroppings of more widespread proliferation in and of themselves also will call forth efforts to reduce the resultant threat of nuclear conflict. But few of the possible measures for mitigating the consequences of proliferation offer a high promise of success, while domestic and international constraints may hinder implementation of even these more limited measures. And the greater the scope, pace, and level of proliferation, the more difficult and complex management efforts will become. Thus, the spiraling risk of regional nuclear conflict will not be entirely offset by these management efforts.

THE GLOBAL SPILLOVERS

While more widespread proliferation most likely will not over-turn the existing structure of world politics, it will adversely af-fect the superpowers, and their relationship, as well as the great powers. The optimism among some analysts about the benign consequences of further proliferation again is likely to be proved wrong.

Limits to Structural Change

The Soviet Union and the United States are involved in nearly all of the regions to which nuclear weapons may spread in the 1980s, frequently supporting opposite sides in long-standing disputes. Neither is likely to sever alliance ties, drop clients and allies, or phase out economic and military involvement after nuclear weapons spread to these regions. In all probability, the leaders of both countries will continue to believe that compel-ling national interests—whether, for example, Western access to Middle East oil, expansion of Soviet power toward the Per-sian Gulf or its containment, the protection of traditional allies, and maintenance of the military balance in East Asia—out-weigh any new or enhanced risks of continuing involvement. Besides, because of the competitive nature of the superpower relationship, officials in each country may be reluctant to dis-engage from these regions in the absence of reciprocal action by the other country lest the opponent be given a "free hand." And an unwillingness to sacrifice past investments made in pursuit of regional influence and military-political advantage is likely to buttress these arguments against disengagement.

It is equally doubtful that more widespread proliferation will lead to a Soviet-U.S. condominium to prevent the further spread of nuclear weapons, ban their use by new nuclear pow-ers, and restore the superpowers' absolute domination of world politics. The competing political, economic, and military inter-ests of the Soviet Union and the United States in regions such as South Asia and the Middle East are likely to take precedence over joint efforts to reduce the risk of local nuclear conflict.

The superpowers' reliance on the nuclear threat in their own defense postures also may constrain joint action, particularly since the threat of escalation to nuclear conflict is critical to NATO's defense posture. The international costs—political, military, and economic—of an attempt to restore superpower domination of regional politics also would be high, and quite possibly thought by U.S. and Soviet leaders to be excessive. For many countries, including U.S. allies in Western Europe, a superpower condominium for nuclear peace would be a grave threat to their current freedom of action. It is also doubtful that the military problems of reasserting control would be manageable at an acceptable cost in light of increased local capabilities for resistance, as exemplified by the Soviet experience in Afghanistan. Moreover, in the Middle East, the economic penalties of intervention, at least for the United States, could be great. And while the domestic political constraints on active interventionism abroad may be less for the Soviet leadership than for U.S. policymakers, in neither country can they be overlooked.

The restoration of a more multipolar global political structure is even less likely to result from the further spread of nuclear weapons. The net impact on superpower strategic dominance of the emergence of a group of lesser nuclear powers will be quite limited. Even the deployment of nuclear forces by Japan and West Germany need not fundamentally upset the existing structure: should the nuclear forces of Japan and West Germany be equivalent to those of France and the United Kingdom, there still would be a considerable gap between the threat they could pose to the superpowers and the threat the superpowers would pose in return. The United States and the Soviet Union also could raise the threshold nuclear capability necessary for Japan or West Germany to mount a serious threat to either of their homelands by renegotiating the 1972 Treaty between the United States of America and the Union of Soviet Socialist Republics on the Limitation of Anti-Ballistic Missile Systems (ABM) to permit Soviet and U.S. deployment of defenses against Japanese or West German ballistic missiles.

Besides, it is quite unlikely in any case that these countries will decide to acquire nuclear weapons.

This conclusion that widespread proliferation will not overturn the existing structure of world politics rests most of all on the assumption that even in that changed environment the leaders of the United States and the Soviet Union will continue to pursue their distinct national interests and objectives, utilizing force or the threat of force and relying on prudence, crisis management, and marginal adjustment to deal with the new risks. However, it is possible that following the use of nuclear weapons by a new nuclear power—especially if that use almost produces a nuclear confrontation between them—the United States and the Soviet Union may be far more ready to negotiate about joint disengagement and other steps to isolate newly nuclear regions. Alternatively, leaders in the Soviet Union and the United States could seek to reassert their countries' capability to dictate the rules of the regional nuclear game. The likelihood of such major adjustments clearly will depend on whether the superpowers' assessment of the direct risks to themselves and of the adequacy of traditional crisis management changes markedly. But particularly in light of the limited success of recent U.S. and Soviet efforts to reach agreement on reciprocal strategic restraints as well as their conflicting global interests, ideologies, and national styles,* even after one or more small-power nuclear exchanges, the two superpowers probably will continue to pursue only prudent ameliorative measures to reduce the risks of competitive involvement in newly nuclearized regions.

Reduced Superpower Freedom of Action

Periodically during the past decades, the United States has intervened militarily in regional confrontations, disputes, and

---

*Soviet stress on unilateral steps rather than mutual restraint to reduce the risk of nuclear war is a specific manifestation of those stylistic differences. See John Erickson, "The Soviet Military System: Doctrine, Technology and 'Style'," in *Soviet Military Power and Performance*, ed. John Erickson and E. J. Feuchtwanger (Great Britain: Archon Books, 1979), pp. 24–29.

limited conflicts outside of the European arena. The decision in 1980 to create the Rapid Deployment Joint Task Force for Middle East and Persian Gulf contingencies reflects a continued willingness to project U.S. power into conflict-prone regions in order to protect U.S. interests, allies, and friends.[11] But the presence of nuclear weapons in some future contingencies will increase the military and political risks of intervention, reducing U.S. freedom of action.

Notwithstanding the threat of U.S. retaliation, nuclear weapons might be used against U.S. intervention forces. A desperate leader, thinking there was nothing left to lose, might launch a nuclear strike against landing troops or close-in offshore naval operations, both of which would be vulnerable to even a few rudimentary nuclear weapons. Or, in the heat of battle, a breakdown of communications could result in the use of nuclear weapons by a lesser nuclear power. Also possible is an unauthorized attack on U.S. forces by the military of a new nuclear power. If needed adaptations of the tactics, training, and structure of these U.S. intervention forces are not made, U.S. intervention could prove very costly, and U.S. forces might even suffer stunning reversals.

Admittedly, U.S. policymakers could launch a limited nuclear strike to disarm the hostile new nuclear power rather than seek to "work around" this regional nuclear threat and risk valuable military assets. But the regional and global political costs to the United States of such a strike are likely to be so high as to make policymakers very hesitant to authorize it.

These heightened risks also are likely to reinforce the lingering, although somewhat muted, national presumption against intervention derived from the Vietnam experience. Consequently, the stakes needed to justify involvement in a newly nuclearized region probably will be greater than in the past. U.S. policymakers may choose not to intervene militarily in some situations where they previously would have acted.

The risks and complexities of military intervention will increase for the Soviet Union as well. In the eyes of a Soviet leadership that has intervened militarily only when the balance of forces appeared clearly favorable, the possible use of nuclear weapons against Soviet troops in a newly nuclearized region could be an excessive risk. To illustrate, Yugoslav deployment of battlefield nuclear weapons might discourage Soviet military action in a future domestic political struggle in Yugoslavia. Similarly, even a slight possibility that Israel or South Africa would use nuclear weapons against Soviet ground or naval forces might help deter Soviet military entanglement in those regions. And the political costs of a nuclear disarming attack on a new nuclear power are likely to appear nearly as excessive to the Soviet Union as to the United States.

The eventual development by a few new nuclear powers of even a limited last-resort capability to threaten the homeland of one or the other superpower with nuclear attack or retaliation also would reduce both Soviet and American freedom of action. For example, should Israel acquire the capability to strike Odessa, Kiev, and Baku, the Soviet leaders might not be as willing to risk direct military involvement in the Middle East to support their Arab clients. Such a capability in Yugoslav or South African hands might have a comparable restraining effect on the Soviets. Or, though less likely, a radical Arab government might threaten to destroy one or more American cities in an attempt to blackmail the United States into not resupplying Israel in the midst of the next Middle East war. Of course, the risk of carrying out such a threat to a superpower would be extraordinary. But neither superpower could ignore the possibility that a leader who thought he had nothing left to lose might do so.

However, this threat of direct attack by a new nuclear power is likely to be greater for the Soviet Union than for the United States. Hardly any of the next countries likely to acquire the bomb will seek to target the U.S. homeland. Moreover, the geographical remoteness of the United States from potentially hostile new nuclear powers in the Middle East and Persian Gulf, combined with the technological backwardness of these countries, makes American cities somewhat less vulnerable than Soviet cities to such a nuclear strike. At least in the 1980s, to attack a U.S. city, Iraq or Libya—the most plausible opponents—probably would either have to smuggle a weapon into

the United States by plane or boat* or use a converted Boeing
707 or 747 registered as a private or corporate jet to deliver a
bomb, counting on subterfuge and the steady decline of U.S.
air defenses[12] to penetrate U.S. airspace. Though possibly feasible, such unconventional modes of attack would be less technically reliable, limited in magnitude, and subject to intercep-
tion by intelligence agencies.

In contrast, Israel, Yugoslavia, and South Korea already possess long-range nuclear-capable aircraft that can reach the Soviet Union and may well be able to slip through Soviet air defenses. South Africa and Israel also are said to be developing
a crude cruise missile that could increase their capability to hit
Soviet cities. Should Japan or West Germany acquire nuclear
weapons, they would have little trouble targeting Soviet cities.
Barring unexpectedly rapid technological progress, the breakdown of current restraints on the sale of cruise missiles and advanced missile guidance systems, or widespread traffic in space-
booster technology and boosters themselves, the United States
will continue to be less vulnerable to nuclear attack by a new
nuclear power than will the Soviet Union—at least into the
1990s.

The constraining effect of more widespread possession of
nuclear weapons of the superpowers should not be exaggerated.[13] The superpowers' readiness and capability to control
events abroad have already been lessened by the decreased legitimacy of using force, rising nationalism, the difficulties of
bringing applicable force to bear in limited disputes, the availability of advanced weapons systems to regional powers, and
the strengthening of countervailing economic instruments of
power.[14] So viewed, the further spread of nuclear weapons
only contributes to a continuing, longer-term relative decline
of superpower freedom of action. Moreover, as long as the two
superpowers are ready to pay the necessary price, they could
preserve a significant gap between their military capabilities

---

*Ton loads of contraband drugs are smuggled routinely into this country by
boat and small plane, and a variety of means of disguising a nuclear weapon to
permit its being included within a larger, more innocent shipment of goods exists. These, however, are best not disclosed, for obvious reasons.

and those of any new medium and lesser nuclear powers, including even Japan and West Germany. Further, in those situations where U.S. or Soviet interests are seen to justify either the military costs of working around lesser nuclear forces or the political costs of suppressing them, the superpowers most probably will be impeded but not prevented from realizing their objectives.

Increased Risk of Superpower Confrontation

Continued U.S. and Soviet pursuit of their respective interests in these newly nuclear conflict-prone regions also will entail acceptance of a higher risk of a U.S.-Soviet political-military confrontation. With the acquisition of nuclear weapons by long-standing regional enemies, there will be many more flashpoints for such a superpower clash. For instance, a preventive attack with conventional weapons, a surprise disarming strike, use of nuclear weapons on the battlefield, nuclear blackmail, a conventional attack backed by the threat of recourse to nuclear weapons, in each case by one superpower's ally against an ally of the other, all could trigger superpower involvement and confrontation. Both the Soviet Union and the United States would be under great pressure to "do something" to help their allies. While aware of the risks, Soviet and U.S. leaders might nonetheless be drawn into the conflict for fear that otherwise their past political, military, and economic investments in the regions would be wasted, their interests sacrificed, and their "reputations for action" ruined. But by responding, the superpower could set in motion an upward spiral of response and counterresponse, of initial entanglement and increased commitment, that may result in a direct confrontation between them.[15]

Though present already, the risk of miscalculation on the part of the two superpowers also may be higher in situations involving newly nuclearized regions—again enhancing the chances of unwanted confrontation. In this new environment, either superpower may modify in unexpected ways its traditional assessment of the stakes, its preferred responses, or its readiness to run risks. Thus, whatever lessons about the other superpower's thinking and responses have been learned from prior regional crises may no longer be fully applicable. And this uncertainty could be most pronounced and most dangerous in the uncharted territory after the next use of nuclear weapons.

Nuclear weapons are likely to increase the tempo of events in regional crises and conflicts, thereby exacerbating further the potential for superpower entanglement and confrontation. For instance, because of the technical deficiencies of their nuclear forces—especially their vulnerability to surprise attack— some, if not many, new nuclear powers will be under considerable pressure to act quickly before their nuclear forces are put out of action. Or, with limited command and control of those nuclear forces if nuclear weapons are used, the pace of the ensuing conflict may be very rapid, with few opportunities to call a halt until the nuclear arsenals of regional opponents are depleted. Finally, because of the destructiveness of nuclear weapons, the threats to the very survival of allies may arise far sooner than when conflicts involved only conventional weapons. Consequently, the superpowers may have to choose quite soon—and probably with even less information than usual— whether to act, and how, or whether to stand aside. And once involved to protect an ally or friend, the superpowers may be overtaken by the heightened tempo of events and pulled into confrontation.

The United States and the Soviet Union undoubtedly will be aware of the risk of confrontation arising out of continued competitive involvement in newly nuclearized regions. Nonetheless, they probably will refuse to sacrifice perceived regional interests, assuming instead that, if needed, they will be able to disengage from a regional conflict before events get out of control. But uncertainty about when to cut losses and disengage as well as the superpowers' declining capability to influence, let alone control, regional events threaten to falsify that assumption. To an unprecedented degree, the actions of regional nuclear powers may force the hand of the superpowers and set in motion a chain of events culminating in a military confrontation that neither the United States nor the Soviet Union wanted but that both are unable to prevent.

Fallout in Western Europe

More widespread proliferation also will reduce the freedom of action of the countries of Western Europe, many of which are involved diplomatically, politically, and, in a few instances, militarily in regions to which nuclear weapons may spread. To illustrate, French advisers are stationed throughout Africa in countries ranging from Gabon to the Ivory Coast, while French troops intervened in Zaire and Chad in the 1970s. There is also a sizable French naval force in the Indian Ocean, and France has strengthened its ties to Saudi Arabia since it assisted the Saudi government in putting down the aborted seizure of the Grand Mosque in Mecca by a fanatical Moslem sect in 1979. Both French and British naval forces have been tacitly cooperating with those of the United States to buttress the Western presence in the Indian Ocean. And, although it has been reluctant to meet U.S. requests for support in the Persian Gulf, West Germany has in the past few years begun to play a more active diplomatic role in the Middle East.

But concern about even indirect entanglement in crises or confrontations that could involve the use of nuclear weapons may make policymakers in these countries more cautious in extending existing political, economic, or military ties. Domestic pressures against heightened involvement could grow as well. Moreover, because of this fear, these policymakers might be even more reluctant to support U.S. military initiatives and may not permit use of facilities and bases on their territories, or agree to reallocate or transship material and supplies, or provide military forces.

As well, these Western European countries might be the targets of nuclear blackmail intended to make them stand aside in such clashes or withdraw previously offered assistance. For example, in the midst of an Arab-Israeli conflict in the late 1980s, Egypt, Iraq, or Libya could anonymously threaten to detonate a nuclear weapon previously smuggled into Portugal unless that country rescinded landing rights at air bases in the Azores for U.S. planes on their way to Israel with needed military equipment. Or West Germany might be the target of such an anonymous nuclear threat in an indirect effort to prevent the United States from shipping military equipment from NATO stocks to the Middle East. Besides, once nuclear weapons are more widely available, it could be quite difficult to distinguish a hoax from a serious threat, and, thus, even a hoax might suffice to disrupt such U.S. operations for a time.

Under certain conditions further proliferation also would increase considerably the cost and difficulty for France and Britain of maintaining a credible nuclear deterrent against the Soviet Union.[16] Confronted by a growing threat to their homelands from new nuclear powers, or believing that such a threat was likely to emerge by the 1990s, the superpowers might renegotiate the 1972 ABM Treaty and deploy ballistic missile defenses.[17] But to counter that change, these medium nuclear powers would have to develop and deploy costly and technically demanding systems able to penetrate those more extensive Soviet missile defenses. Failure to do so would lead to the increasing obsolescence of the French and British nuclear forces.*

DOMESTIC POLITICAL REPERCUSSIONS

Nuclear-Armed Terrorists, Irredentists, and Separatists

A considerably greater risk that terrorist and dissident groups will gain access to nuclear weapons will be another adverse consequence of the further spread of nuclear weapons.[18] As more countries seek to acquire a nuclear weapons option by initiating sensitive reprocessing or enrichment activities, or set up actual weapons programs, the number of sites from which these groups could steal nuclear weapons material for a bomb will increase.[19] The ensuing transportation of such material between a growing number of sites will further increase the risk of theft. Once a group possesses nuclear weapons material, the

---

*China's nuclear force would be similarly threatened with obsolescence by a Soviet missile defense capability.

technical hurdles of processing that material and fabricating a nuclear weapon still would have to be overcome, but at least for some groups these difficulties would not be insurmountable. More important, a subnational group might opt for stealing the bomb itself, taking advantage of the probably less-than-adequate physical security measures of some new nuclear forces.

Hit-and-run clandestine terrorist groups, such as the Japanese Red Army, extreme left-wing Palestinian factions, the Italian Red Brigade, the Irish Republican Army (IRA), or successors to the Baader-Meinhof gang, may well regard a nuclear weapon as a means of extorting money or political concessions from a government, much as taking hostages is now.[20] The countries of Western Europe, Japan, and the United States will be especially vulnerable to terrorist threats or attack because of their open societies. A group such as the IRA, claiming to represent a legitimate alternative government and dependent on popular support, might stop short of carrying out a nuclear threat even if its demands were not met. But members of the more radical and nihilistic fringe movements, such as the successors to the Baader-Meinhof gang and the Japanese Red Army, might think otherwise. To them, carrying out the threat might appear justified as a means of bringing down corrupt bourgeois society in a spasm of violence. Or, in the eyes of the most extreme Palestinian groups, use of a nuclear weapon might be thought justified as a way of mortally wounding Israel. Yet again, with the police closing in on them, these more radical, isolated terrorists could conclude that, since all was lost, it would be preferable to fall in a nuclear *Götterdämmerung*. Such a decision would be consistent with the near-suicide mentality shown in some past terrorist actions.[21]

In contrast, the theft and threatened use of nuclear weapons may not appear a worthwhile tactic to a group such as the Palestinian Liberation Organization (PLO). Even though the PLO's freedom of action has been reduced by the Lebanese civil war, it still controls territory, administers to its refugee population, has a military force, and has been recognized by international bodies and foreign governments. Rather than enhancing the PLO's claim that it is a legitimate government in exile, possession of a few stolen nuclear weapons could have the opposite effect. Theft of nuclear weapons would reinforce the PLO's reputation for extremism and unwillingness to accept minimal norms of international behavior and would make it harder for those Western European governments moving closer to the PLO's position on the Middle East to sustain that shift. Besides, should Israeli intelligence manage to locate these nuclear weapons, pressures to carry out a preventive strike, disregarding the risk of Soviet counteraction, would be intense. If Israel could not locate the nuclear weapons but knew that the PLO had them, the result is not likely to be Israeli acceptance of the need for a Palestinian state but Israeli unwillingness to compromise on that PLO demand. On balance, therefore, the costs to the PLO of stealing nuclear weapons appear to outweigh the benefits. Still, that conclusion reflects a Western weighing of costs and gains, which may prove as unfounded in this instance of Middle East maneuverings as it has on earlier occasions.

Separatist movements such as the Kurds or Arabs in Iran, the Baluchis in Pakistan, the Bengalis in India, the Moslems in the Philippines, or even the Basques in Spain might be more inclined to steal and threaten to use a nuclear weapon. For example, a separatist Baluchi movement might threaten to use stolen Pakistani nuclear weapons if the Pakistani central government mounted a new military campaign to restore its authority. Though extreme, such a threat would not be out of line with the bitter fighting so characteristic of these struggles for greater autonomy. Fearful of the consequences of cracking down on the separatists and under international and domestic pressure to find a "reasonable" settlement, the central government might come to terms with that group. Conversely, the central government could conclude that the costs of yielding to the separatists' demands were so great that it had no choice but to strike back, even using its own nuclear weapons against those of the separatists. But with little to lose, the Baluchis—and other separatist groups in other countries—might be ready to take that chance.[22]

The Nuclear Coup d'Etat

In the 1980s and early 1990s, politically unstable new nuclear powers—such as Argentina, Brazil, Chile, Egypt, Iran, Iraq, Libya, Nigeria, Pakistan, South Korea, and Syria—might be vulnerable to nuclear coups d'etat.[23] Particularly if the balance of political and military power between the rebels and the government were unclear, control of nuclear weapons—as compelling a symbol and instrument of national power as control of the airport, capital city, or radio and television stations—could greatly enhance the rebels' bargaining position. Control of nuclear weapons would change the psychological climate and afford rebel groups a means not only of demoralizing opponents but also of rallying supporters. The specter of nuclear destruction—should the situation get out of hand—quite possibly might lead civilian and military fence-sitters to come out in favor of a coup and even change the minds of some anti-coup forces. Moreover, just a few nuclear weapons in rebel hands could suffice to deter attack against them, assuming that the government was both unwilling to overwhelm the rebels with conventional force lest they retaliate with nuclear weapons and reluctant to use nuclear weapons first on its own territory in a surprise disarming attack. Consequently, more so than in past coups, efforts to dislodge such rebels would remain a test of wills and bargaining strategy. Nevertheless, nuclear weapons might be employed, either intentionally, by accident, or out of contempt and hatred.

Already on at least one occasion during the first decades of the nuclear age, access to nuclear weapons has figured in a domestic political upheaval. In April 1961, French army forces stationed in Algiers rebelled, demanding that the government in Paris reverse its decision to grant independence to Algeria. At the time, French scientists were preparing to test a nuclear weapon at the French Saharan test site at Reganne, Algeria, not too far from Algiers. Noting the proximity of the rebellion, the scientists called on the general in charge at Reganne to authorize an immediate test and thus avoid the possibility that the nuclear device would be seized by the rebel troops and used

for bargaining leverage.[24] Three days after the outbreak of the
revolt, the order to detonate the device came directly from
French President de Gaulle; there was no attempt to undertake
precise experiments, only to use up all the available fissionable
material.

Further, the prospect that a country's nuclear weapons
might fall into the "wrong hands" could even provoke outside
military intervention. A neighboring country, for example,
might launch a disarming attack or, if time and the situation
permitted, try to transport the entire nuclear arsenal out of the
country.* Direct military support for a government confronted
by a coup, or support for a countercoup to evict the radicals
after a coup occurred, also might be offered. Thus, the nuclear
coup d'etat is yet another flash point for regional conflict and
possible nuclear escalation.

The Corrosion of Liberal Democracies?

At least some of the measures required to deal with the threats
of clandestine nuclear attack—whether from a terrorist group
or a new nuclear power—and of nuclear black marketing will
be in tension with or in outright violation of the civil liberties
procedures and underlying values of Western liberal democracies.[25] For example, to hinder clandestine efforts by a subnational group or a new nuclear power to introduce, move, or
make ready a preplaced nuclear device, strong restrictions on
movements into, out of, and within the United States and other
open societies are likely to be adopted. An intelligence warning
of an attempt to smuggle a bomb into the southwestern United
States might be followed by a temporary ban on private flying
in that region. New laws might be enacted as well to control
more tightly the registration of aliens currently in the United
States with ties to other countries, to facilitate the expulsion of
both legal and illegal aliens with possible connections to terror-

*In a somewhat comparable situation, just before the fall of South Vietnam
in 1975, the United States removed and flew out of the country the nuclear
materials in South Vietnam's one operating research reactor to prevent their
seizure by North Vietnam.

ist or black market organizations, and to regulate more stringently the movement of aliens within this country.

Because of the stakes, there will be strong pressures to circumvent or set aside—in the United States and elsewhere—various constitutional and legal restrictions on invasions of privacy or other traditional civil liberties.[26] Unauthorized, warrantless emergency searches based on skimpy evidence or tips might be made. Or broad neighborhood—even city-wide—searches may become legitimate in these instances, although existing laws in many countries, particularly the Fourth and Fourteenth Amendments in the United States, prohibit searches without specific definition of the site and evidence sought. The use of informants, warrantless or illegal wiretaps, and the secret detention and questioning of suspects for days or even weeks might follow, all motivated by the need to acquire information as fast as possible. Highly coercive interrogation methods, ranging from painless but effective truth serum drugs to more extreme forms of physical deprivation and psychological disorientation, even to more brutal forms of torture, are not precluded. Further, limited press censorship to avert public panic and resultant pressures to make concessions, or simply to avoid tipping one's hand to the opponent, might be instituted.

Some countries acknowledged to be liberal democracies already have adopted some of these measures to deal with conventional terrorist threats. The British Parliament's Northern Ireland Act of 1973, for example, allows for detention without warrant, while in West Germany there are restrictions on the right to counsel for members of terrorist groups.[27] Even in the United States there have been past abridgments of civil liberties reluctantly justified as necessary to preserve the overall fabric and underlying democratic values. Thus, faced with an extreme nuclear threat, a future U.S. president may argue successfully, as did Lincoln during the Civil War when he suspended the writ of habeas corpus, "Are all the laws, but one, to go unexecuted, and the government itself to go to pieces, lest that one be violated?"[28]

However, it may prove possible to contain this challenge to

liberal democratic procedures and values. Within the United States, both rigorous administrative supervision of any emergency measures and strict judicial review after the fact would help prevent those measures from spilling over their boundaries and corrupting procedures in other areas of law enforcement. Authorizing legislation and official policy statements also could stress the extraordinary character of those restrictions as a response to an exceptional threat while reemphasizing the more basic American belief in the worth, dignity, and sanctity of the individual that underlies respect for particular civil liberties.

But if the frequency of proliferation-related threats grows, and if violations of traditional civil liberties cease to be isolated occurrences, it will become more difficult to check this corrosion of liberal democracy here and elsewhere. For that reason, as well, concern about the many adverse consequences of increasingly widespread nuclear weapons proliferation is well founded.

5 • CHECKING THE BOMB'S SPREAD

Virtually all thinking about the problem of nuclear weapons proliferation has been couched in terms of preventing even one additional country from joining the nuclear club. This definition of the proliferation policy agenda clearly is no longer adequate. The emergence of some new nuclear powers is unavoidable. But it may be possible to contain the nuclear genie by simultaneously adopting measures to slow proliferation's pace and check its scope, to minimize the adverse repercussions of the initial outcroppings of further proliferation in the years ahead, and to mitigate the local and global consequences of the nuclearization of conflict-prone regions.

The chances for holding the line at no more than a continuation of the first decades' pattern of slow and limited proliferation still can be influenced significantly by the policies pursued by the United States and other countries. Arguments that if one or two additional countries acquire nuclear weapons there will be a flood of decisions to "go nuclear" exaggerate some countries' incentives to acquire the bomb, ignore others' disincentives, and neglect various proliferation firebreaks. Moreover, even if a somewhat more extensive erosion of the first decades' pattern cannot be avoided, it still will be important to contain the spread of nuclear weapons as much as possible, for such efforts can mitigate the grave problems of living in a world of more nuclear weapons states. Thus, checking proliferation's scope and pace has to remain the bedrock of even this broadened policy approach; nevertheless, the nonproliferation initiatives pursued by the Carter Administration must be revamped and supplemented.

BUTTRESSING THE TECHNICAL HURDLES

The Carter Administration focused almost exclusively on measures to slow the erosion of technical constraints confronting countries seeking to acquire nuclear weapons. A continuing effort to induce countries to defer the reprocessing and commercial use of plutonium was at the core of its nonproliferation offensive. Diplomatic persuasion, the example set by the United States in deferring such activities, and especially the condition—provided in most U.S. agreements for nuclear cooperation with other countries—that the prior consent of the United States be obtained before U.S.-origin or enriched nuclear fuel could be reprocessed all were employed in an effort to restructure the rules and practices governing the international use of nuclear energy.

But many countries, including some close allies, considered the Carter Administration's policies to be an attempt to force their legitimate nuclear energy activities into an American-made procrustean bed.[1] As a consequence, they repeatedly rejected the U.S. position. Japan, for example, began to operate a pilot-scale reprocessing facility in 1978, created a company to build a second reprocessing facility, sent fuel to France and the United Kingdom for reprocessing, and went ahead with research and development on the generation of electricity with plutonium-fueled fast breeder nuclear reactors.[2] Cooperating with Belgium and the Netherlands, West Germany also began building an experimental fast breeder reactor and refused to set aside the option of reprocessing spent fuel from power reactors as a means of reducing the amount of nuclear waste.[3] France not only reprocessed fuel for a growing number of countries but also took a major step toward the commercial use of plutonium by beginning construction of the Super-Phenix fast breeder reactor.[4] Many developing countries expressed dissatisfaction with what they saw as an attempt to deny them access to necessary peaceful nuclear technology, while a few of

the more advanced of this group such as Argentina, Yugoslavia, and Brazil, talked of establishing their own suppliers'
group so that they could provide each other with sensitive nuclear energy technology.[5]

Nevertheless, it is necessary not to overreact to this failure of
the Carter Administration's measures by returning to a much
more laissez-faire approach to other countries' nuclear energy
activities. Steps still are warranted to mitigate the proliferation
threat represented by widespread national reprocessing and/or
enrichment. But rather than futilely trying to dictate how other
countries use nuclear energy, a more flexible approach that
seeks to foster a set of presumptions or ways of thinking about
sensitive nuclear activities is demanded.

Influencing Global Nuclear Energy Activities

As a start, while acknowledging that some spent fuel reprocessing is unavoidable, the United States still should strive to limit
the number of countries engaged in that activity. It also should
try to promote an understanding that the separated plutonium
be used only for experimental purposes or as fuel for breeder
reactors but not be used (recycled) as fuel for light-water power
reactors (LWRs). To reduce the availability of large quantities
of readily accessible weapons-grade nuclear material, the
United States also needs to strengthen the presumption against
economically premature research, development, or deployment of breeder reactors, quite rightly contending that such
activities are justified only in countries with large electric grids
and advanced nuclear industries. Similarly, to avoid the emergence of many small, pilot-scale enrichment facilities capable of
producing limited quantities of weapons-grade uranium, such
as that in South Africa, the notion that enrichment facilities
should be deployed only when justifiable on economic and
technical grounds by the country's other nuclear energy activities ought to be promoted by the United States. Finally, a presumption of heightened multinational or international involvement in the development, ownership, and operation of new
reprocessing and enrichment ventures should be supported. If
successfully disseminated and institutionalized, these presump-

tions would reduce significantly the risk of proliferation inherent in the global use of nuclear energy, but they would stop short of trying to dictate other countries' legitimate energy choices.

Economic, technical, and political considerations already are pushing countries in the direction suggested by these presumptions. For example, the global slowdown in the deployment of nuclear reactors has reduced the demand for uranium and undermined the economic justification for the early commercial use of plutonium.[6] That same global slowdown has led to a surplus of capacity for enriching uranium, thereby checking pressures to acquire national enrichment facilities.[7] Similarly, continuing high capital and operating costs projected for breeder reactors have slowed French plans for these reactors' rapid commercialization.[8] Technical problems with handling bulk quantities of spent fuel from power reactors have impeded Japan's plans for commercial reprocessing, and they are likely to have the same impact elsewhere.[9] And domestic political opposition in West Germany and Japan has adversely affected the pace of both countries' fast breeder reactor programs.[10]

Immediate steps are needed, nonetheless, to encourage directly these presumptions about how best to use nuclear energy and to foster their implementation. Bilateral discussions as well as U.S. presentations in multilateral forums could play a useful role. And because other countries still pay some attention to the U.S. nuclear energy program, a decision by the Reagan Administration not to recycle reprocessed plutonium could also affect those countries' perception of desirable future plutonium uses. Most important, after several years of international discussions and negotiations, institutional building blocks of a new global approach to using nuclear energy must be put into place.

Security of nuclear supply remains a lingering, if somewhat reduced, concern of many countries. An international network of nuclear fuel assurances needs finally to be established to guarantee an adequate supply of nuclear fuel for countries that would otherwise feel pressured to shift to the commercial use of plutonium or to develop a national enrichment capacity as a defense against politically or economically motivated supply disruptions.[11] All countries that purchase nuclear power plants could be contractually guaranteed at the time of sale a supply of nuclear fuel for the life of the reactor—assuming there is no violation of any legally binding nonproliferation obligation. If that guarantee were not sufficiently credible to some buyers, those countries also might be offered "cross guarantees," whereby other fuel suppliers agreed to supply needed fuel if the primary supplier failed to do so for any reason other than violation of a nonproliferation obligation. In some regions, stockpiles of nuclear fuel might even be established as a further reassurance. Such a multilayered set of fuel assurances could be gradually implemented, responding to the specific requirements of particular countries and taking into account the readiness of other nuclear suppliers to cooperate.

Though thus far impeded by practical problems, alternatives to national storage of spent fuel are required. They would help reduce pressures for reprocessing that stem from the problem of managing nuclear waste and the fact that existing national storage capacity has been exhausted in some countries.[12] (In a few countries, provision for the reprocessing of spent fuel already is a condition for acquiring a license to operate a nuclear power plant.) To that end, the United States ought to make a final effort to reach agreement with Japan and South Korea on a regional spent fuel storage center for East Asia. A serious attempt also should be made to overcome the domestic political hurdles to carrying out the offer made by President Carter in 1977 to take back some spent fuel from foreign countries when the United States had originally supplied the fuel. In both cases, however, the prospects for success are mixed.

Another useful and perhaps more realizable step would be the establishment of multilateral institutions to provide developing countries possessing advanced nuclear programs with access to sophisticated nuclear fuel cycle research and development—particularly of fast breeder reactors, which many of these countries are already beginning to think of as a possible next step in their nuclear development. Conceivably, arrange-

Estimating the Risk of Nuclear Weapons Proliferation*

○ Motivation   ● Capability

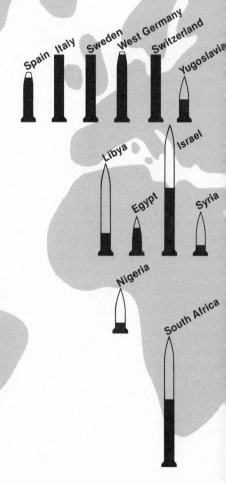
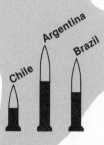

Spain Italy Sweden West Germany Switzerland Yugoslavia

Libya Israel

Egypt Syria

Mexico

Nigeria

South Africa

Chile Argentina Brazil

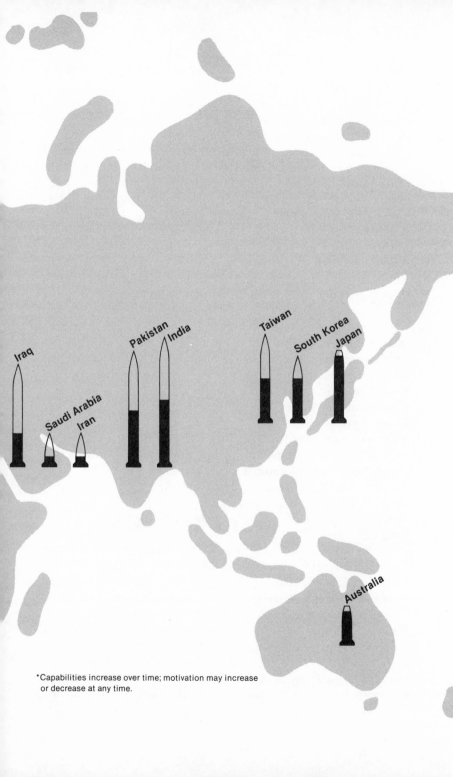

*Capabilities increase over time; motivation may increase
 or decrease at any time.

ments might be made for scientists from those developing countries to participate in research programs under way in industrial countries. Or, one or more industrial countries with sophisticated nuclear programs might help several advanced developing countries establish a joint research facility. Still another possibility would be participation in an international energy research and development center, perhaps under International Atomic Energy Agency auspices. Such multilateral institutions would provide countries with rapidly developing nuclear energy programs—such as South Korea, Taiwan, and Argentina—with a means of acquiring information necessary for planning those programs so that they might not launch their own pilot-scale research efforts. Although the level of technological sophistication and hands-on experience gained from participation in a multilateral venture would be higher than if these countries had to proceed on their own, that potential proliferation drawback is outweighed by the advantages of limiting the spread of pilot-scale sensitive facilities and research programs and by the greater opportunity to influence these countries' nuclear energy planning. In addition, access to multilateral research and development of advanced nuclear energy systems could help reduce the prevalent perception of technological discrimination against developing countries that could lead several countries to withdraw from the NPT, thereby seriously weakening the treaty. And if that perception could be reduced, developing countries would more readily accept international safeguards as a condition for access to peaceful nuclear technology.

As the opportunity arises, the United States also should support multilateral involvement—whether ownership or management—in new enrichment ventures. For example, the United States could respond favorably to Australia's request for American participation in its planned first enrichment facility.[13] Australia also should be urged to accept participation by one or more advanced developing countries, such as South Korea. Efforts also are needed to encourage the internationalization—perhaps in the form of equity participation by several countries in return for a guaranteed share of the output—of the commercial-scale Japanese enrichment facility now under development. And, similar involvement of foreign capital might still be sought for new U.S. enrichment plants. Greater acceptance of multilateral involvement in enrichment activities would not only slow the spread of small enrichment facilities, it would also set a precedent for increasing internationalization of other sensitive nuclear activities. Thus, its long-term significance would be considerable.

Finally, it is especially important for the United States to play a more active role in establishing procedures for international storage of separated plutonium (IPS). Supporting IPS would not be easy for the United States because it would legitimize and routinize access to plutonium by some developing countries with advanced nuclear programs. But the United States alone cannot counter the strong international support for IPS. A U.S. posture of diplomatic stonewalling may actually be counterproductive because it will sacrifice the opportunity to influence the activities of those countries that will go ahead with plans to reprocess. Consequently, it appears advisable to seek the best deal possible, trading active U.S. participation for a tighter system of controls on separated plutonium. In particular, U.S. negotiators might seek, for example, to foster agreement that plutonium will be released from international stockpiles only for experimental uses or as fuel for breeders; to tighten the safeguards and conditions governing the uses of plutonium; to limit the number of plutonium storage sites and the stock of plutonium permissible at sites in nonnuclear weapons states; and to provide for multilateral sanctions in response to any violation of the IPS agreement—including a legal commitment by nuclear suppliers to cut off automatically any further nuclear transactions.

Greater U.S. flexibility in dealings on nuclear energy matters with the countries of Western Europe and with Japan could help generate support for these initiatives. For example, if the United States were to cooperate with rather than oppose Japan's efforts to build a second reprocessing facility, Japan might be more willing in return to go along with tighter controls in an IPS scheme or agree to participate in an Asian spent fuel storage center. Or, a readiness on the part of the United States to approve generically—as opposed to on a case-by-case basis—the reprocessing of U.S.-originated or -enriched spent fuel by France and the United Kingdom might be exchanged for these countries' support for the U.S. position on plutonium storage. Greater U.S. toleration of the legitimate nuclear energy activities of these allies also should bolster efforts to reach international agreement on several remaining nuclear suppliers issues.

Closing Exports Loopholes

The first signs of what threatens to become a lessening of nuclear exports restraint are already visible. Italy's sale to Iraq of a hot-cell for reprocessing plutonium is just one example. It is particularly important that the Reagan Administration take steps to check this trend lest the erosion of technical constraints accelerate, adversely affecting proliferation's scope and pace. To avoid any misperception of its position, it would be helpful if the new Administration made clear its support for the nuclear suppliers agreements of the Ford-Carter years and strenuously used its diplomatic influence with the countries of Western Europe and with Japan to oppose further backsliding. Particularly since the new Mitterand government in France is likely to be more restrained in its nuclear exports, renewed efforts to reach agreement with the major nuclear suppliers on requiring full-scope safeguards as a condition of supply could be another step worth taking. Safeguards on all nuclear facilities would make it more difficult to misuse or divert material from those facilities, while acquisition of nuclear weapons would thereafter violate a legally binding nonproliferation obligation. Consequently, the risks and costs of nuclear weapons acquisition would be greater because of the increased likelihood that painful international economic, nuclear energy, political, and other sanctions would be imposed.

Additional measures are needed to tighten controls over gray market nuclear materials and components. These items, including those with a legitimate civilian use but also usable for military purposes—such as frequency inverters and other electrical components—have played a critical role in Pakistan's quest for the bomb and could be equally significant to the nuclear weapons activities of other countries. Current bilateral diplomatic efforts to convince other nuclear suppliers to tighten their controls on the export of such components and materials should be supplemented by attempts to agree on more formal limitations concerning gray area exports. In many instances, additions could be made to the "trigger list" of items to be exported only under certain conditions and with safeguards. Or ad hoc agreements might be sought from time to time with individual nuclear suppliers to proscribe the export of specific items, for example, electrical components for Pakistan's enrichment facility. And in tightening these controls, attention increasingly will have to be paid not only to components of use in detonating a crude bomb but also to components of use in reaching higher levels of nuclear weapons capability.

Discussions with other suppliers to improve procedures for sharing intelligence data on suspected gray and black marketing are in order as well. The lack of early detection of Pakistan's circumvention of exports controls clearly suggests that improvement is needed. With more timely intelligence, it might be possible to prevent companies from selling or delivering gray market components and materials, to arrest black marketeers, to thwart possible theft of material, or even to intercept black market transfers.

Further, the United States should encourage—using whatever mixture of bilateral exchanges and multilateral forums appears suitable—at least de facto acceptance of minimal nonproliferation standards by third-tier nuclear suppliers such as Argentina, Brazil, and India, whose nuclear activities are not covered by the NPT or by existing nuclear exports controls and procedures. Should these countries' nuclear supply activities require U.S. approval—that is, if they involve retransfers of nuclear material supplied originally by this country—it probably would be advisable to provide such approval as long as these activities are consistent in practice with nonproliferation objectives and norms. Trying to get these third-tier suppliers to declare explicitly that they will honor those norms probably would only result in bitter and self-defeating disputes. But while these suppliers are likely to be unwilling to agree in principle to restraints on their nuclear exports, they may be ready to observe them in their day-to-day actions.

Nevertheless, taking these steps to buttress the technical barriers to nuclear weapons acquisition will not be enough to check the pace and scope of proliferation. For a growing number of countries, overcoming the technical obstacles to acquiring at least simple fission bombs is becoming ever easier. Thus, additional initiatives, for the most part neglected during the Carter years, must be taken to influence countries' incentives and disincentives to acquire nuclear weapons.

A SANCTIONS STRATEGY

Fear of adverse foreign reaction is a significant disincentive to the acquisition of nuclear weapons. Consequently, U.S. government officials, outside observers, academics, and spokesmen for the nuclear industry have all called for the strengthening of sanctions as part of U.S. nonproliferation policy.* Recent U.S. legislation has taken some steps in that direction.

The Foreign Assistance Act of 1961, as amended in 1976 and 1977, now prohibits military and virtually all types of economic assistance to countries delivering or receiving enrichment or reprocessing equipment, materials, or technology.[14] Also amended in 1977, the Export-Import Bank Act of 1945 precludes financial support for transactions with any nonnuclear weapons state that engages in nuclear weapons-related activities.[15] Most comprehensive, the Nuclear Non-Proliferation Act of 1978, which amends the Atomic Energy Act of 1954, requires halting nuclear exports to nonnuclear weapons

---

*Margaret Doxey defines sanctions as "conformity-defending instruments relating to behavior which is expected by custom or required by law." They are to serve "accepted norms of international conduct," not the particular policies of a given nation. See Margaret Doxey, *Economic Sanctions and International Enforcement* (London: Oxford University Press, 1971), pp. 1, 14.

Sanctions may include economic, financial, political, energy, diplomatic, or military measures.

states found by the President to have engaged in activities involving nuclear weapons material or of direct significance to the manufacture of nuclear explosive devices; to have detonated a nuclear device; or to have violated, terminated, or abrogated IAEA safeguards. It also calls for termination of nuclear exports to countries assisting, encouraging, or inducing nonnuclear weapons states to engage in activities related to the manufacture or acquisition of nuclear weapons. Recognizing that pursuit of other U.S. foreign policy and security interests might require a more flexible posture, both the Foreign Assistance Act of 1961 and the Nuclear Non-Proliferation Act of 1978 allow for presidential waiver of these sanctions under specified conditions subject to congressional override.[16]

There also are provisions for international sanctions against the acquisition of nuclear weapons. Article XII(c) of the IAEA statute empowers its Board of Governors to curtail or suspend nuclear assistance (regardless of its supplier), to call for the return of nuclear materials or equipment, and to suspend any country from membership in the IAEA if that country violates international safeguards.[17] Supplementing that IAEA provision, the London Nuclear Suppliers Group has agreed to consult in the event of a suspected violation of the London Guidelines or of IAEA safeguards to determine whether a violation has actually occurred and, if so, how to respond. They have agreed that their response "could include the termination of nuclear transfers to [the] recipient."[18]

But experience in implementing these sanctions has been discouraging. In the first significant test case, U.S. military and economic assistance to Pakistan was suspended in late 1976 and again in early 1979 because of that country's acquisition of reprocessing and then enrichment technology. Neither step induced Pakistan to end those activities. Then, in 1981, the Reagan Administration, as part of its attempt to buttress the Western security position in the Persian Gulf, sought to waive those restrictions and institute a five-year program of economic and military aid. Both the initial failure of sanctions to reverse Pakistan's policy and the later U.S. decision to provide substantial assistance to that country suggest that more thinking is re-

quired about the purposes, risks, and strategy of sanctions in
order to come to a more realistic understanding of what they
may be able to achieve.

What For?

The reasons for threatening to impose sanctions should be dis-
tinguished from those for actually carrying out that threat.
The primary purpose of threatening sanctions is to deter a
country from undertaking nuclear weapons-related activities.
The fear of adverse economic, political, security, nuclear energy, or other punitive measures might help tip the balance of
disincentives and incentives in favor of not "going nuclear."

Even if the threat of sanctions is unable to realize that objective, it still may suffice to change the scope and characteristics
of a country's nuclear weapons activities, which in itself may be
beneficial. For example, fearful of the threat of sanctions, a
country might purposefully steer clear of violating international safeguards in its efforts to obtain nuclear explosive material, and if it is a party to the NPT, it might first withdraw
from the Treaty in order not to violate its legal obligations. As
a result, the adverse impact on efforts to check proliferation
would be less severe. Or, to minimize the threat of sanctions, a
country might only acquire a limited covert capability—just as
Israel and South Africa appear to have done—rather than deploy a full-fledged overt nuclear force, while still other countries might go no further than a single detonation of a "peaceful nuclear explosive." So holding down the level of nuclear
weapons activity could provide neighboring countries with an
incentive to stop at only matching programs. At the same time
it would minimize the adverse impact of that activity on fears
of runaway proliferation. Other potential benefits of avoiding
deployment of full-fledged overt nuclear forces include fewer
command-and-control problems (since covert forces would be
smaller and probably centrally stored), a lower risk of accidents
than with operationally deployed nuclear weapons, and, consequently, a reduced chance of regional nuclear conflict.

If the threat of sanctions fails to deter an outright nuclear
weapons program, punitive measures then might be imposed

in a last-resort attempt to coerce the target country to reverse or at least halt those activities. Indeed, reversing a country's action after the fact has been the traditional, and virtually never realized, objective of recourse to sanctions.

Moreover, whether or not the actual imposition of sanctions is unlikely to bring about a reversal of policy, it may still be worthwhile to impose sanctions in order to influence other countries' perceptions of *their* freedom to pursue nuclear weapons. If the threatened sanctions are not imposed, these onlookers most likely will reassess downward the costs and risks of launching their own nuclear weapons programs. Conversely, carrying out the threat of sanctions may induce other countries to assess quite carefully their proliferation incentives and disincentives, and at least some of them may discover themselves to be far more vulnerable than the target country to the sanctions just imposed—or to other punitive measures.

Depending on the specific situation, the imposition of sanctions also may be necessary to demonstrate a readiness to stand behind nonproliferation agreements, norms, and expectations. Failure to impose sanctions in the event of violations of the Nuclear Suppliers Guidelines will undermine the resolve of major nuclear suppliers to observe exports restraints as well as the willingness of third-tier suppliers to adhere to minimal nonproliferation guidelines. Similarly, the credibility and operating effectiveness of the safeguards system requires the imposition of sanctions in the event of a safeguards violation, lest other countries reassess the risk of such violation and IAEA morale plummet. More broadly, imposing the threatened sanctions could help counter the belief that widespread proliferation is becoming unavoidable and challenge the legitimacy of acquiring nuclear weapons.

Measures Available

A variety of measures can be threatened or invoked by the United States, preferably with the support of allies, against countries setting out to acquire nuclear weapons. (These measures are summarized in Table 1.) There is no one universally applicable sanction; different countries are more or less vul-

Table 1. Unilateral or Multilateral Sanctions

- Termination of nuclear assistance, exports, and cooperation
- Complete or selective embargo on imports and exports
- Cutoff or reduction of bilateral economic assistance
- Slowdown or cutoff of World Bank loans
- Cutoff of U.S. Ex-Im Bank loans and comparable support
- Ban on private investment and lending
- Freeze on financial assets abroad
- Refusal to refinance outstanding debt
- Termination of airline landing rights
- Expulsion of foreign students working toward technical, scientific, engineering, and similar degrees as well as postdegree trainees
- Ban on provision of economic, technical, managerial, and other assistance
- Reassessment of alliance relationship
- Withdrawal of troops and other reductions of alliance commitment
- Termination of security ties
- Breaking of diplomatic relations and withdrawal of political support
- Embargo on arms sales or transfers
- Expulsion from appropriate international agencies

nerable to different measures. Most countries, however, are vulnerable to one, if not more, of the sanctions listed.

The termination of supplies of nuclear fuels, spare parts, and associated assistance would seriously affect the economic well-being and energy plans of quite a few countries. Taiwan and South Korea, for example, are each planning to depend increasingly on nuclear power for the generation of electricity,[19] while Argentina and Brazil also have serious nuclear power programs and Pakistan continues to be interested in setting one up.

Several countries with potential proliferation incentives are vulnerable to trade sanctions because of their dependence on trade with a limited number of countries. About half of South Korean trade in 1979, for example, was with the United States and Japan, and that foreign trade accounted for more than 50 percent of the South Korean gross national product.[20] Taiwan's economic position is much the same. Other countries, while not vulnerable to an across-the-board trade embargo, depend on particular commodities or services that can only be acquired through imports. South Africa, for example, is extremely dependent on imported oil,[21] while Libya relies heavily on U.S. technology and U.S. and Western personnel for continued operation and expansion of its oil fields.[22]

Still other countries are dependent on foreign economic assistance, including multilateral aid from the World Bank, as a source of needed investment capital. Sanctions that cut off or reduce that foreign assistance, say to India or Pakistan, in all probability would lead to significant economic dislocation, particularly in economic sectors targeted for emphasis in these countries' development programs.[23] Similarly, the economic well-being of Pakistan and Brazil would be seriously hurt if the United States and a handful of other creditor countries refused to reschedule debt repayments.[24] Alternatively, because of their dependence on foreign capital flows and, to a lesser degree, direct investment, a multilateral ban on private lending and direct investment to such advanced developing countries as South Korea, Taiwan, or Brazil could be equally disruptive.[25] And with their considerable deposits in foreign banks, countries such as Libya, India, and Iraq would be economically shaken by a freeze on those financial assets abroad.[26]

For Israel, South Africa, and Taiwan, loss or simply reduction of U.S. and, in some cases, Western political and diplomatic support would exact a considerable price for "going nuclear." Without that support, these countries would be virtually isolated diplomatically and more vulnerable to outside pressure or even military assault. Such isolation might even lead to a collapse of domestic morale.

The threatened termination of arms sales is another more political and security-related sanction. The security of Israel, South Korea, and Taiwan, for example, depends on the availability of sophisticated U.S. conventional arms and technology. Iraq and Libya rely heavily on arms supplied from France and the Soviet Union, while India depends on supplies from the Soviet Union and Britain. Further, several of the most sensitive prospective new nuclear powers are heavily dependent on a continuing—if sometimes shrinking—American security guarantee. Any reassessment of that guarantee would undermine the defense postures of South Korea and Taiwan in Asia, Israel and Egypt in the Middle East, and to a degree, Pakistan in South Asia.

The importance of multilateral as opposed to unilateral U.S. action will vary from measure to measure and case to case. More often than not, to inflict painful economic sanctions the cooperation of other countries will be needed to buttress unilateral U.S. economic measures. The number of countries whose support will be required—and their identity—varies, but rarely is more than three or four. By contrast, the United States acting alone would be able in some cases to impose costly nuclear program-related sanctions or to withdraw political-security support.

The difficulties of putting together a sanctions coalition will depend partly on the number of other countries whose support is required. Their readiness to cooperate most likely will be greater in the event of violations of legal nonproliferation obligations or safeguards than if considerable ambiguity surrounds the target's nuclear weapons activities. Another significant consideration will be the possible adverse economic or political effects on the coalition partners themselves of carrying out threatened sanctions. Thus, while a limited number of countries control a variety of coercive levers that, if used, would impose considerable cost on the target countries, they sometimes may be reluctant to use those levers, thereby lessening both the threat and the impact of sanctions.

How Effective Are They?

The threat of sanctions has been able to tip the balance against acquisition of nuclear weapons in several situations where perceived incentives still were limited and a firm commitment to acquire the bomb had yet to be made. For example, in late 1975 and early 1976, according to officials in Seoul, the United States made the "strongest possible representations" to dissuade South Korea from purchasing a French reprocessing plant and otherwise continuing its efforts to acquire nuclear explosive material. Not only future sales of nuclear fuel and technology but also the total alliance relationship were used as levers.[27] The threat of U.S. sanctions was successful and South Korea—deciding that its efforts to hedge against future U.S. disengagement might end up provoking that U.S. action— shelved its nascent nuclear weapons program.

Still another example occurred in the fall and winter of 1976, when the United States exerted considerable pressure on Taiwan to discontinue secret reprocessing of spent fuel. Taiwan's dependence on American-supplied nuclear fuel was emphasized, while the importance of political and security ties to the United States also entered into the discussions between officials of these two countries. Again, the implied threat of sanctions was effective. In November 1976, despite consistent denial that it had been reprocessing, Taiwan provided explicit assurances that it would not reprocess spent fuel, that it would dismantle and seal existing reprocessing facilities, and that it would allow an IAEA-U.S. inspection team to verify that these steps had been taken.[28]

In still other cases, the threat of sanctions has led countries to modify the pace or characteristics of their nuclear weapons activities. For example, in early August 1977, the Soviet government informed the United States that it had detected preparations for a South African nuclear weapons test. U.S. intelligence satellites confirmed the existence of an underground nuclear test site in the Kalahari Desert.[29] During the next two weeks, the United States, France, West Germany, Britain, and the Soviet Union mounted a strenuous diplomatic and political offensive to head off the nuclear weapons test. At stake, they made clear, were continued South African access to civilian nuclear technology and facilities, South Africa's diplomatic ties, and the readiness of Western governments to block adoption of stringent U.N. economic sanctions advocated by black African states.[30] On August 22, the South African government yielded to that pressure and provided assurances that it did not intend to and would not develop nuclear explosives and that no nuclear test would be undertaken in South Africa then or in the future.[31] Nonetheless, the detection of an apparent nuclear test off South Africa on September 22, 1979 suggests that South Africa's program may only have been slowed, not shut down.

Israel's decision to stop short of overt deployment of nuclear weapons also undoubtedly has been affected by concern about an adverse foreign—particularly American—reaction. And the increasingly negative international diplomatic reaction that followed India's detonation of a "peaceful nuclear device" in May 1974 certainly contributed to that country's decision to defer follow-up nuclear testing and, in the words of one Indian official, "to let the political dust settle."*

However, if the threat of sanctions does not suffice to deter a country from initiating nuclear weapons activities, it is most unlikely that their actual imposition will force the target country to halt or reverse those activities. Indeed, over the past decades, recourse to sanctions has consistently failed to coerce countries to change their policies.

In 1935, for example, when Italy invaded Ethiopia, the League of Nations tried to force Italy to withdraw by imposing sanctions that included a ban on arms shipments, restrictions on financial dealings, and a partial trade embargo that excluded shipments of oil or other essential raw materials such as iron, steel, and coal. Though these sanctions proved highly inconvenient, Italy refused to withdraw from Ethiopia. In fact, the sanctions were counterproductive, as they rallied nationalistic support for Mussolini, his regime, and the Ethiopian war.[32]

British and United Nations economic sanctions against Rhodesia in the mid-1960s were unable to shake the Smith regime's refusal to negotiate transition to black majority rule. As a result of those sanctions, popular support for Smith increased. Further, with the assistance of the Salazar government in Portugal and the Afrikaner regime in South Africa, such effective means to circumvent sanctions were developed that the Rho-

---

*The lack of keen interest of India's military high command in nuclear weapons, limits on Pakistan's capacity to respond to India's nuclear activity, and the change of government in March 1977 also contributed.

desian economy grew at a rate of 10 percent each year through the early 1970s. What led eventually to the end of white minority rule in Rhodesia and the creation in 1980 of black-governed Zimbabwe were the emergence of a serious guerrilla struggle and the withdrawal of South African support from the Smith government.[33]

While U.S.-imposed economic sanctions against Cuba in the early 1960s contributed to that country's economic decline in the decade that followed, Soviet assistance and a fundamental shift in Cuban trading and financial arrangements saved Cuba from economic collapse. At no time did the Castro government face a serious domestic threat due to the economic repercussions of the sanctions, nor did it consider modifying its policies.[34]

More recently, the 1979 U.S. cutoff of economic and military assistance to Pakistan was unable to reverse that country's quest for nuclear explosive material. Likewise, the imposition of U.S. trade and financial sanctions against Iran—with eventual qualified allied support—in response to the seizure of U.S. hostages in November 1979 heightened the ongoing deterioration of the Iranian economy; but at least until Iraq's invasion of Iran in September 1980, which made Iran's access to its frozen assets more important in order to pay for military supplies, these sanctions had little, if any, impact on the policies of the Ayatollah Khomeini and his supporters. And, even then, the hostages' release was mostly due to the changing requirements of Iranian revolutionary politics.[35]

Actual recourse, as opposed to the threat of sanctions, failed in these and still other cases partly because their impact, though painful, was bearable by the target country. Quite frequently, countries with considerable leverage refused to participate in sanctions or the participating countries shield away from adopting the most coercive measures, while external support sometimes offset the sanctions adopted. Moreover, the target countries were firmly committed to the particular policies that had triggered sanctions. But, most significantly, in each of these cases, the perceived costs of yielding to demands for a change in policy were considerably higher than the costs of the sanctions themselves.[36] Domestic political support, if not the very domestic social and political order; acceptance of lesser international status; reduced influence on critical foreign policy issues; and even national survival were variously seen to be at stake.

Far less data are available on the indirect or symbolic effects of carrying out threatened sanctions. However, there is some evidence that the U.S. decision in October 1978 to impose a total trade ban against the Ugandan regime of Idi Amin Dada helped to persuade Tanzania's President Julius Nyerere that the international community would not oppose the military action eventually taken by Tanzania to overthrow Amin and encouraged Amin's opponents to concert their efforts against him.[37] It also has been suggested that the imposition of sanctions against Cuba demonstrated U.S. resolve to oppose communism in Latin America.[38] Although there is only limited empirical information available, the theoretical argument that recourse to sanctions may influence onlookers' perceptions of their freedom of action while demonstrating support for nonproliferation agreements, norms, and expectations seems sufficiently forceful to justify its acceptance in evaluating what the threat or imposition of sanctions can be realistically expected to achieve.

The Risks of Sanctions

There are, nevertheless, risks incurred if sanctions are adopted. Among them is the possibility that the particular measures will be in tension with other foreign, domestic, or economic objectives. For example, the cutoff of U.S. economic and military assistance to Pakistan in 1979 because of its acquisition of enrichment technology clashed with U.S. efforts to strengthen ties with Pakistan following the Soviet invasion of Afghanistan. Similarly, in the event of, say, a Pakistani nuclear test, reimposition of now-waived restrictions on such aid would run counter to the Reagan Administration's attempt to forge a new containment posture in Southwest Asia. Or, U.S. support for sanctions after an Israeli nuclear test would risk a domestic political outcry in the United States, while to go along with an embargo on trade with South Africa, should it detonate a nuclear explosive device, could significantly reduce U.S. supplies of certain strategic minerals such as chromium, manganese, and platinum, which are predominantly produced by South Africa.

Another frequently cited risk entailed in imposing sanctions is the possibility that the target country will adopt economic countermeasures in retaliation. Though likely to have less of an impact in times of glut than scarcity, oil suppliers, such as Libya and Iraq, for example, could cut off exports to countries imposing or joining a sanctions coalition. Or, in response to trade, investment, or financial sanctions, a country could freeze payments on its foreign debt, compensating in the short run for the lost access to foreign loans and economic aid by shutting off the outflow of capital. In addition, existing foreign investment could be nationalized, foreign workers deported, exports of critical raw materials embargoed, or remittances due private foreign investors blocked. While the available potential economic countermeasures—and their impact—will vary from case to case, few countries lack some form of retaliation. However, they quite frequently may be reluctant to use the more extreme of these measures, such as nationalizing investment or repudiating debt, lest it damage their own economic well-being.

In contrast, political countermeasures may have fewer adverse repercussions for the target country and thus they will be easier to adopt. Libya, India, Argentina, and Iraq, for example, all could respond to sanctions by moving closer to the Soviet Union.

Going beyond such standard retaliatory countermeasures, a new nuclear power could threaten to provide other countries with nuclear weapons-related assistance—helping to design and construct indigenous facilities for producing nuclear explosive material, supplying components, or even aiding in the design and testing of a nuclear weapon. In apparent response to public discussions of sanctions against India following its nuclear test in May 1974, India's government pointed out just how responsible its nuclear exports policy had been. While India most likely would have stopped short of carrying out that implied threat to export nuclear weapons assistance, not least because it would have entailed serious risks, in the next decade an "international pariah" might regard such a step as a legitimate act of self-defense or reprisal.

Still another danger is that if a country had yet to decide to launch a full-fledged nuclear weapons program, imposing sanctions in response to its initial nuclear weapons activities may crystallize that very decision—for the costs of sanctions might be sufficient to rule out a middle course, forcing a choice between desisting or seeking the perceived benefits of going full ahead. Costly sanctions also could strengthen the hand of the pro-nuclear leadership by rallying domestic support, thus enabling the government to pursue a more ambitious nuclear weapons program than had previously been planned. Such an expanded nuclear weapons program in turn would exacerbate the adverse impact on still other countries' proliferation calculations—just the opposite result than originally intended.

Because of these risks, a sanctions strategy must provide some flexibility to permit policymakers to make adjustments to particular situations. But excessive flexibility, leading repeatedly to decisions not to impose sanctions, must be avoided. Such inaction would increase many countries' perceptions of their freedom of action and undermine nonproliferation efforts.

To balance the requirement of flexibility with that of ensuring some expectation of a punitive response, a sanctions strategy needs to contain a mix of automatic and optional measures. Thus, it ought to be declared that, as required by the Nuclear Nonproliferation Act of 1978, the United States at the very least will automatically cut off American nuclear assistance and exports to a country that violates a legal nonproliferation obligation and to any nonnuclear weapons state that detonates a nuclear explosive device. Concomitantly, the United States should encourage the presumption that there will be not only no waiver of other sanctions contained in U.S. legislation but that additional punitive measures not mandated by law might be adopted as well. Further, presidential policy statements, international agreements, legislative resolutions, and quiet diplomacy all might be used by the United States to create the presumption that other, even if legal, nuclear weapons activities short of open testing also might trigger a broad range of sanctions.

Adding Some Muscle

It is not too early for U.S. policymakers to begin thinking about possible responses to a variety of nuclear activities by specific countries. While any detailed plans would be kept confidential, an occasional leak that such contingency planning was being undertaken could increase the credibility of the threat of sanctions. U.S. intelligence efforts to monitor the activities of prospective nuclear powers in order to provide earlier detection of violations and more response time might be augmented as well.

The United States also should ask the major nuclear suppliers to agree to a rule of collective responsibility for responding to violations of nuclear obligations. Extending the suppliers' present commitment not to "act in a manner that could prejudice any measure that may be adopted by other suppliers . . . ,"[39] this new rule would call for each major supplier to treat a violation of another supplier's agreement as a violation against itself and to take appropriate action. In addition, provision that any violation of safeguards or other legally binding nonproliferation obligations would bar a country from access to peaceful nuclear commerce ought to be sought. However, rather than reconvening the London Nuclear Suppliers Group—which the developing countries oppose as an instrument of domination—more ad hoc bilateral or more formal multilateral agreements that extend suppliers' restraint should be pursued.

Particularly where technical ambiguity about whether an activity was proscribed could lead to controversy and inaction, discussion needs to be encouraged among the major suppliers to define more precisely what would constitute a violation of the Nuclear Suppliers' Guidelines. Laying the foundation for later timely reaction, the United States also should promote continuing consideration by these countries of possible sanctions measures and the specific situations that could warrant their imposition. Such discussions might even lead to suppliers' agreement on procedures either to neutralize or share the economic costs and other burdens of retaliatory countermeasures by the target country. This, too, will not only make it easier to put together a sanctions coalition but also will enhance the credibility of the threat of sanctions. For without such procedures, countries may be reluctant—as they have been in the past—to carry out sanctions.

While the proposed sanctions strategy will help enhance proliferation disincentives, measures to check the growth of incentives for acquiring nuclear weapons must be undertaken simultaneously, for once incentives reach a certain intensity and bureaucratic and public support coalesces in favor of "going nuclear," even the threat of costly sanctions will be outweighed by the perceived payoff of a nuclear weapons capability. And should a growing number of countries disregard the threat of sanctions, that threat will lose credibility to a point where it will be insufficient even to influence countries without pressing motivations to acquire nuclear weapons. Besides, unless steps are taken to provide insofar as possible an alternative means of meeting a prospective nuclear power's security concerns, threatening sanctions to keep it from "going nuclear" will be of questionable legitimacy. Consequently, both the threat and use of sanctions may fail to receive needed domestic and international support while engendering even more widespread discontent among the nuclear have-nots about discrimination against them. For that reason as well, sanctions are a necessary but not sufficient part of continuing U.S. nonproliferation efforts.

CHECKING PROLIFERATION INCENTIVES

Both broad global initiatives designed to reduce all countries' proliferation incentives and more country-specific measures to achieve that objective have been proposed. The most widely favored global initiatives, however, probably will have only a marginal impact and for the most part they appear unrealizable.

A Pledge of Nonuse

Throughout the 1960s and 1970s, many countries with only conventional military forces urged the nuclear powers to pledge not to use nuclear weapons against nonnuclear weapons states. In 1978, Secretary of State Cyrus Vance responded that:

> [T]he United States will not use nuclear weapons against any non-nuclear-weapons state party to the NPT or any comparable internationally binding commitment not to acquire nuclear explosive devices, except in the case of an attack on the United States, its territories or armed forces, or its allies, by such a state allied to a nuclear-weapons state or associated with a nuclear-weapons state in carrying out or sustaining the attack.[40]

But this broad pledge did not satisfy many nonnuclear weapons countries. Those that were not party to the NPT or to a comparable agreement were excluded from the guarantee, while there were still conditions under which even parties to such agreements could be threatened by the use of nuclear weapons.

As long as NATO's defense posture and protection of the vital U.S. interest in European independence and stability continue to depend on the threatened use of nuclear weapons against the Soviet Union and its nonnuclear allies,[41] the scope of any further U.S. nonuse guarantee is clearly limited. Nonetheless, the United States might state that it will not use nuclear weapons against a nonnuclear weapons state with the single exception of countries participating in a Warsaw Pact attack on Western Europe—explicitly ruling out the use of nuclear weapons against virtually all nonnuclear states, whether parties to the NPT or not. This guarantee would preclude U.S. use of nuclear weapons in Korea, but there is considerable doubt about the efficacy of employing nuclear weapons there.[42] And even though the guarantee probably would heighten South Korea's insecurity and thus its incentives to acquire the bomb, that adverse effect could be limited by maintaining a U.S. troop presence and offset by the threat of sanctions.

Such a qualified pledge or, for that matter, even an unqualified U.S. pledge not to use nuclear weapons against nonnuclear weapons states—supported by comparable pledges by the four other acknowledged nuclear powers—probably would have only a very marginal effect on reducing the proliferation incentives of those countries that could "go nuclear" in the next decade. Though references to the superpower threat are a useful rhetorical gambit for proponents of nuclear weapons, fear of a U.S. or a Soviet nuclear attack is unlikely to be significant in the nuclear weapons calculations of such countries as Pakistan and India, Iraq and Israel, South Korea and Taiwan, South Africa and Nigeria, or Argentina and Brazil. The determining considerations are more likely to be concern about a hostile newly nuclearized neighbor, erosion of the local conventional military balance, pursuit of regional or global prestige and influence, domestic politics, and bureaucratic and scientific momentum. Where concern about the superpowers exists, it derives mainly from fear of intervention with conventional forces—a fear that would not be reduced substantially by a joint superpower nuclear nonuse pledge.

While the immediate direct impact of a U.S. pledge not to use nuclear weapons against nonnuclear countries except in the event of a NATO-Warsaw Pact clash is marginal, such a qualified guarantee could prove far more beneficial in the long run. For example, that pledge might assuage some of the nuclear have-nots' resentment of the unwillingness of the nuclear haves to accept any restrictions on their own nuclear weapons activities—a resentment that could influence the have-nots' later nuclear decision making. Debate over providing the nonuse guarantee, likely to raise questions about why it is in the U.S. interest to take steps to check proliferation, might also increase the acceptability of U.S. security guarantees for individual countries. If the other acknowledged nuclear powers made comparable nonuse pledges, it would establish a precedent or rule that new nuclear powers might be induced to honor, thus conceivably reducing somewhat both security-related and domestic political pressures to "go nuclear" in nonnuclear neighbors. Further, should the superpowers, say in the wake of a future proliferation shock, assert their joint intention to enforce it, a qualified rule proscribing the use of nuclear weapons against nonnuclear countries could help restrict recourse to nuclear threats in a world of more widespread proliferation.

Nuclear De-emphasis

A second frequently proposed broad global initiative for checking proliferation incentives is to reduce the role of nuclear weapons in world politics—"nuclear de-emphasis" for short.[43] Large-scale reductions of Soviet and American strategic forces, a comprehensive ban on the testing of nuclear weapons (CTB), and a U.S. declaration that it will not use nuclear weapons first even against a nuclear power, all are part of nuclear de-emphasis.

Underlying this proposal is the valid contention that other countries' perceptions of the utility of nuclear weapons are affected to some degree by the nuclear weapons policies, postures, and doctrines of the United States and the Soviet Union. The continuing buildup of Soviet strategic forces and the Reagan Administration's concern with restoring the U.S. strategic capability lest a dangerous military and political imbalance result reflect a belief that nuclear weapons can significantly influence the outcome of crises and confrontations. Opposition within the United States to a CTB teaches that nuclear weapons may be politically and militarily useful and therefore desirable. And continued U.S. reliance on the threat to use nuclear weapons first if needed in a conventional conflict in Europe or South Korea conveys the message that for countries confronting a conventionally superior opponent nuclear weapons are a valuable deterrent. (By contrast, the nonuse of nuclear weapons by the United States in either the Korean or Vietnam wars—neither, for example, to destroy Chinese forces in Korea in November 1950 nor to close off the Mu Gia pass on the Ho Chi Minh trail during the Vietnam War—suggests that the payoffs of acquiring these weapons are more limited.)

There is some evidence to support the assumption that the superpowers' policies have affected the thinking of countries that might "go nuclear." For example, a few Indian and Pakistani officials have argued that much as the threat of nuclear conflict eventually led both East and West to accept the status quo in a divided Europe, it would do so on the borders between India and China and India and Pakistan.[44] Similarly, one argument sometimes made for Israeli overt acquisition of nuclear weapons is that their deployment will lead to a stable Middle East balance of terror resembling that between the superpowers.[45] References to the value of "mini-nukes" to buttress Yugoslavia's defense undoubtedly were stimulated by earlier debate within the U.S. defense community about the utility of such weapons.[46]

Further, the U.S. and Soviet readiness to make nuclear threats and failure to "slow the nuclear arms race" as required by Article VI of the NPT[47] also will figure in other countries' domestic debates about "going nuclear." References to these "sins" of the superpowers provide a convenient debating point for the proponents of acquiring nuclear weapons, help put their opponents on the defensive, and deflect assessment of the risks and costs of deploying a nuclear force.

Nevertheless, a superpower posture of nuclear de-emphasis will not decisively affect, for example, the strategic thinking of Pakistan, India, Iraq, Taiwan, South Africa, Israel, Argentina, or Brazil. These countries' particular political and military environments will be a far more potent determinant of their leaders' perceptions of the utility of acquiring nuclear weapons. Moreover, while rhetorical references to superpower reliance on nuclear weapons can be expected, the outcome of the policy debate will mostly turn—judging from past experience and limited discussion already under way—on arguments about the implications of changing regional conventional military balances; whether future security requirements could be met more effectively by nuclear deterrence; the purported global status and prestige to be derived from "going nuclear"; the need to match traditional rivals' activities; the risk of sanctions; the danger of a nuclear coup d'etat; and, to a lesser degree, the costs of nuclear weapons acquisition. Not least of all, whether the next country to use nuclear weapons achieves its military and political objectives or suffers a disastrous defeat will heavily influence these perceptions of the benefits and costs of acquiring a nuclear arsenal.

Furthermore, the marginal positive impact of nuclear de-emphasis on perceptions of nuclear weapons' utility and on the internal policy debate in nonnuclear countries may exact too great a price, for nuclear de-emphasis will heighten appreciably the likelihood of a West German and, to a lesser degree, a Japanese decision to acquire nuclear weapons. West German confidence in the NATO alliance and the U.S. nuclear umbrella will remain critical to minimizing its incentives to acquire nuclear weapons. A no-first-use pledge by the United States would undermine that confidence because the threat of such use is at the heart of NATO's posture for deterring a Soviet conventional attack on Western Europe. And even if matched by the Soviet Union, a large reduction of the U.S. strategic arsenal probably would be perceived by West German leaders as indicative of a greatly reduced readiness on the part of the United States to use its strategic forces on behalf of its allies in Western Europe.[48] Marked reductions of U.S. strategic capabilities also will increase Japan's fears of a withdrawal of the American nuclear umbrella and generate pressures for a reassessment of that country's nonnuclear status. And if West Germany or Japan "goes nuclear," the prevailing pattern of slow and limited proliferation easily could break down altogether.

Besides, a policy of nuclear de-emphasis raises questions that extend far beyond the realm of proliferation policy to encompass assumptions about the meaning of the continuing across-the-board Soviet military buildup; the role of U.S. strategic forces in deterring Soviet activism and buttressing the U.S. bargaining position; the risks of large nuclear force reductions; the balance of conventional forces in Europe; the limits to détente; and so on. These other calculations have consistently taken precedence over any putative nonproliferation payoffs in strategic decision making, and there is little reason, given the preceding analysis, to change that ranking. While it is useful and desirable to bear those possible payoffs in mind when one thinks about future Strategic Arms Limitations Talks

(SALT) agreements, a CTB, or a no-first-use policy, the ultimate decision on any such measures justifiably will turn on calculations that transcend proliferation policy itself.

A Comprehensive Test Ban Treaty

A comprehensive test ban treaty has most often been discussed
as part of a program of nuclear de-emphasis intended to lower
widely held perceptions of nuclear weapons' utility. But it also
has been argued[49] that if most or all of the existing nuclear
powers would agree to and honor a CTB, considerable pressures might be brought to bear on such potential new nuclear
powers as India and Pakistan; Israel, Iraq, and Libya; South
Africa; Argentina and Brazil; and Taiwan and South Korea to
do the same. Adherence to a CTB would restrain the nuclear
weapons activities of the signatories themselves. Of equal importance, it would lessen neighboring countries' incentives to
launch overt nuclear weapons programs by constraining the
signatories' activities, by providing a legal affirmation of the
signatories' intentions not to test, and by reducing mutual suspicion and uncertainty.

There are reasons to doubt, however, whether prospective
nuclear powers in South Asia, the Middle East, and elsewhere
will become parties to a CTB treaty. While occasional Indian
statements have implied support for a CTB, for example, they
made superpower strategic arms reductions a condition for
signing and, in any case, were issued by the former Desai government.[50] Besides, if China refuses to adhere to such a treaty,
India's concern about China's nuclear weapons capability, reinforced by a deep desire for equal status with China in Asia, is
likely to provide a compelling motive for its nonadherence. In
turn, China's reluctance to accept a permanent position of
marked strategic inferiority to the Soviet Union raises doubts
about its readiness to participate in a CTB. And even if China
does sign, the unwillingness of India's leaders to accept a sec-
ond-class political and military position could lead to Indian
abstention. Without India's adherence, of course, Pakistan will
not sign. Even if India consented to a CTB treaty, Pakistan still
might keep open its nuclear weapons option as a hedge against

a steadily improving Indian conventional military capability and because of domestic political pressures.

Much the same situation holds true in other regions. The diplomatic and military benefits of Israel's ambiguous nuclear status, the unwillingness of its leaders to mortgage their future security options, and the likely nonadherence of neighboring Arab countries all suggest that Israel will not sign a CTB treaty. It is hard to believe that, given U.S. domestic politics, sufficient pressure could be brought to bear by the United States to change that decision. Iraq and Libya are just as unlikely to sign because it would limit their attempt to alter the Middle East political status quo. It is equally implausible that South Africa's leaders would restrict their freedom of action in a hostile international environment by adhering to such a treaty. Even in Latin America, Argentinian and Brazilian insistence on keeping open their nuclear weapons options— partly because of internal political and bureaucratic pressures— may well outweigh the potential benefit of a CTB treaty in reducing residual mutual suspicion. Only South Korea and Taiwan appear likely to sign a CTB to avoid alienating their American protector.

Because of these doubts about the willingness of potential new nuclear powers to adhere to a CTB treaty, assessment of how strongly to pursue this measure on nonproliferation grounds must rest on its contribution to a posture of nuclear de-emphasis. As such, it suffers from that posture's limits, suggesting again that the greatest part of the burden of checking the growth of proliferation incentives will have to be borne by more country-specific measures.

The Alliance Connection

Preserving the strength and reliability of American alliances and more informal security ties is especially significant for containing proliferation since their erosion would spur the growth in quite a few countries of security-related proliferation incentives. The continued presence of American ground forces in South Korea, for example, buttresses that country's security, thus weakening the case for a South Korean nuclear force. It also helps head off any reinvigoration of Japanese concern about the reliability of the U.S. guarantee—a concern that was prevalent following the U.S. defeat in Vietnam and again when Carter announced the withdrawal of troops from South Korea—thereby reducing as well the likelihood of Japan's rethinking its nuclear weapons abstinence. Also, preserving NATO's credibility and avoiding West German reassessment of its nonnuclear weapons status depend in part on the presence of U.S. troops in Europe and also on the perceived sufficiency of U.S. strategic forces backstopping the alliance.

Perceptions of the reliability of existing U.S. alliances can be eroded, however, by weak or inappropriate U.S. responses in international crises and confrontations. In the mid- to late-1970s, the aborted U.S. involvement in Angola and the limited U.S. response to Cuban and Soviet military intervention in the Horn of Africa created uncertainty in Saudi Arabia, Israel, pre-Khomeini Iran, and elsewhere about the wisdom of relying on the United States.[51] Saudi concern was further increased by the United States' inability to foresee or respond effectively to the 1979 collapse of the Shah's regime in Iran.[52] And part of the U.S. response to reassure the Saudis—sales of advanced jet aircraft as well as deployment of U.S. airborne warning planes at the beginning of the Iraq-Iran war—added to Israel's doubts.

Individual American foreign policy initiatives also influence other countries' assessments of the costs and benefits of depending on the United States. A U.S. readiness not to talk directly with the PLO or to push too hard for a "comprehensive" Middle East settlement long has been a critical weather vane, in Israeli eyes, of the health of the U.S.-Israeli relationship. Or, if Japan is pushed too far, U.S. attempts to induce that country to revise its trade policies can embitter relations and undermine Japanese confidence in the United States. Even proliferation policy can undermine those perceptions: the Carter changes in U.S. nuclear exports policy affected not only the credibility of the United States as a civilian nuclear supplier but also broader assessments of American reliability.

In light of conflicting U.S. policy interests, however, it may not always be possible to preserve the reliability of existing American alliances. As the severing of formal U.S. diplomatic and military ties with Taiwan made clear, pursuit of other military, foreign, or international economic policy initiatives on occasion will weaken the alliance network. Even so, it may be possible to continue less formal security ties and thus limit the adverse proliferation effects of those overriding policy objectives.

But not all of the countries that could seek nuclear weapons in the mid- to late-1980s are currently tied to the United States by formal alliances. For those prospective new nuclear powers whose proliferation incentives and disincentives are fairly evenly balanced, a security guarantee from the United States could be sufficiently credible to tip the scale against acquiring nuclear weapons. Had the United States pledged to support Pakistan against Indian nuclear blackmail or attack in response to Prime Minister Bhutto's request for such a guarantee following India's 1974 test, there is good reason to believe that Pakistan's ensuing march to the bomb would have been either shelved or greatly constrained.[53] And even as late as 1981, the readiness of the Reagan Administration to provide economic and military assistance to buttress Pakistan's security still could have a comparable effect. Should Iraq or Libya acquire nuclear weapons, strengthening informal U.S. security ties to Egypt and Saudi Arabia—say by a presidential pledge supported by Congress, showing the flag, or even by signing a more formal agreement—might lessen the proliferation multiplier effect. New security ties to Nigeria might sever an African proliferation chain that could follow confirmation of South Africa's possession of nuclear weapons.

While memories of the Vietnam War have faded, the American public still may be chary of providing additional security guarantees—a reluctance that probably would be reinforced by the knowledge that providing guarantees to some of the countries most in need of more rigorous security assurances may clash with other American foreign, security, and domestic policy interests. Stronger U.S. security ties to Pakistan, for example, clash with efforts to improve relations with India; closer ties to Taiwan could preclude further normalization of relations with China; and assurances to Egypt or Saudi Arabia might meet with domestic resistance because of their implications for Israel. Besides, new or stronger security ties would heighten the risk of U.S. involvement in a regional conflict.

Nevertheless, there are likely to be situations—such as if Saudi Arabia is threatened by Iraqi nuclear blackmail—where the arguments favoring more extensive U.S. security ties will be strong enough to overcome both practical difficulties and domestic opposition. Moreover, in the aftermath of one or more proliferation shocks—say, three or four countries' decisions in rapid succession to deploy nuclear weapons or several withdrawals from the NPT—U.S. reluctance to make new security commitments probably would decline, for it will become obvious to many that the alternative to greater U.S. involvement is a more extensive erosion, if not a breakdown, of the first decades' proliferation pattern as well as a heightened risk of nuclear conflict.

Conventional Arms Transfers

Though more controversial because of its risks, the selective transfer of conventional arms—including precision-guided antitank and antiaircraft weapons as well as precision-guided bombs, area denial munitions, and scatterable land mines—also could contribute to containing the growth of some countries' incentives for acquiring nuclear weapons.[54] Fear of conventional military attack by a more powerful local rival or of an unexpected setback on the battlefield is a significant incentive for acquiring nuclear weapons that might be checked by possession of advanced conventional arms capable of buttressing defenses.[55] Conventional arms transfers also can reinforce and strengthen—though not substitute for—U.S. security assurances. Continuing access to advanced U.S. weapons systems is a symbol of the vitality of the U.S. security connection to South Korea and to Israel, for example, as well as of the American readiness to honor the residual security pledge of the Taiwan Relations Act. And since the transfer of conventional arms frequently has been a sign of the importance that the United

States attaches to relations with the recipient country, it thus may provide some of the status that might otherwise be sought through nuclear weapons acquisition. For instance, the 1981 sale of advanced AWACs aircraft to Saudi Arabia demonstrated that the United States regarded that country as an important friend and believed that the opinions of its leaders had to be taken seriously. Or, within Latin America, the purchase of high-performance aircraft is less a response to military and security requirements than a precondition for regional status and influence.[56]

Arms transfers, however, will check only certain incentives for acquiring nuclear weapons, leaving others intact.[57] Selective arms transfers are unlikely to change decisions to "go nuclear" driven by domestic political and bureaucratic pressures and by fear of nuclear blackmail or attack by a neighboring country, let alone by a desire to support adventuresome foreign policies, to threaten neighboring countries, or to backstop conventional uses of force.

Using arms transfers as a nonproliferation measure also entails risks whose magnitude will depend on the specific case. The likelihood and destructiveness of local conflicts could be augmented. American relations with other countries in the region might be damaged. Or, on occasion, transferring advanced arms to reduce one country's incentives to "go nuclear" might increase those of its neighbor, which then might seek nuclear weapons to reassert its military superiority and claim to regional influence. The recipient even could reassess its earlier decision not to acquire nuclear weapons once it had received the advanced conventional arms, thereby strengthening its overall military position.

These risks need not rule out the use of conventional arms transfers for nonproliferation purposes. Rather, they indicate that a careful country-by-country, region-by-region assessment of the magnitude of the risks and of the extent to which selective arms transfers appear capable of influencing the recipient, as well as of the likely consequences of the overt spread of nuclear weapons to the region in question, must be made. There will be instances justifying acceptance of such risks, not least because the destructiveness of a nuclear war between small powers could greatly exceed that of conventional conflicts.

Confidence-Building Measures

Since uncertainty about the nuclear weapons intentions of neighboring countries may fuel proliferation, confidence-building measures to reduce those uncertainties or to slow their growth in regions where one or more countries already have acquired nuclear weapons also are useful policy measures. Indeed, one of the main purposes of the NPT is precisely to reduce those uncertainties by providing a means whereby a country can attest to its peaceful intentions. Questions do arise, of course, about the seriousness of some of the parties' commitment to the Treaty and about their readiness to honor its restrictions should their political and security environment change greatly. And Article X provides for a right of withdrawal if a country's "supreme national interests are threatened." But particularly where adherence to the Treaty has followed a nationwide and bureaucratic debate, it probably comprises a significant impediment to future nuclear weapons activities and should serve to reassure nervous neighbors. The likelihood, however, of convincing more countries—especially such holdouts as Brazil, Argentina, Israel, India, and South Africa—to become parties to the NPT is slim. They will be reluctant to give up their nuclear weapons option and to accept the diminished status of permanent nuclear have-nots. With discontent about the Treaty growing—stimulated in part by claims of insufficient access to advanced nuclear technology—heading off possible withdrawals from the NPT will demand equal, if not greater, attention than recruiting new members.

Gaining acceptance of full-scope safeguards (covering all of a country's civilian nuclear activities) is another confidence-building measure.* While India, Israel, and South Africa are unlikely to agree to them for reasons of status and security,

---

*Safeguards on all peaceful nuclear activities already are required of NPT parties.

other countries—perhaps Argentina or Brazil—just might be induced to accept them as the price of access to advanced civilian nuclear technology. A hard-line U.S. bargaining position and renewed efforts to get the other major nuclear suppliers—including the possibly more cooperative French government of President Mitterand—to agree to full-scope safeguards as a condition of supply could pay off.

Nuclear weapons–free zones (NFZs)—whereby the countries of a region renounce acquisition of nuclear weapons and accept external verification of their intentions—also may be useful in reducing mutual uncertainty and suspicion that can result in one country's attempt to acquire the bomb before its traditional rival has done so.[58] The Latin American Treaty of Tlatelolco, not yet fully implemented, establishes such an NFZ; initiated and negotiated by the countries of that region themselves, that Treaty has proved a more acceptable confidence-building measure than the NPT.

Unless other overriding U.S. interests are at stake, U.S. support for NFZs appears justified by their potential payoff in checking incentives to acquire nuclear weapons. At least the United States should ratify Protocol I of the Treaty of Tlatelolco which applies the ban on nuclear weapons to territory in the NFZ under U.S. control, for example, the Virgin Islands. By failing to ratify that protocol, the United States provides a convenient excuse for Argentina and Brazil not to implement the Treaty. Quiet diplomatic initiatives also might be taken to urge Argentina to carry out its pledge in 1977 to ratify Tlatelolco and to encourage discussions between Israel and Egypt to narrow the gap between their respective plans for a Middle East NFZ. The United States also could declare its readiness to abide by any African NFZ, while using its diplomatic influence to foster talks on its creation. In contrast, because an Indian Ocean NFZ would hinder the ability of the United States to protect vital Western interests in the Persian Gulf—if only because aircraft carriers that routinely carry nuclear weapons would be prohibited from that region—a more cautious posture is warranted in that case.

U.S. support alone cannot assure the success of efforts to

create an NFZ in Latin America or elsewhere. Creation of NFZs is infeasible without the willingness of countries to cede their nuclear weapons options. But without U.S. support—if only in the form of a readiness to abide by the NFZ's require- ments and to seek to induce other nuclear powers to do so as well—it will be extremely difficult to establish an NFZ even in regions, such as Latin America, where the countries themselves agree on its benefits in principle but are hesitant to take that last step to implement their agreement.

Foreign Policy as Nonproliferation Policy

Finally, the tone of American foreign policy can critically affect proliferation incentives. While negotiating on issues from trade to security, the United States can either convey a sense that it is seriously examining and heeding the views of countries such as Argentina, Brazil, and India with ambitions for regional and global influence or engender resentment that can feed and reinforce their incentives to "go nuclear." In turn, doubts abroad about the U.S. ability to deal effectively with the in- creasingly complex problems of world politics have been grow- ing, thereby adding to many countries' uncertainty about their political and security milieu. As a result, reversal of that image of an indecisive and weak America could lessen pressures on some countries to acquire a nuclear weapons option, let alone the bomb, as a hedge against later threats.

Though subject to a high risk of failure, American-sup- ported diplomatic initiatives to alleviate or resolve local con- flicts that often lie at the root of security-related quests for nu- clear weapons also would be useful. For instance, agreement on a comprehensive Middle East peace settlement, encouraged by U.S. mediation and a readiness to use its influence with Is- rael, as well as, perhaps, guaranteed by a U.S. military pres- ence in the region, could prove far more critical than any one other policy measure in heading off the overt nuclearization of the Middle East. Or, U.S. influence in China might be used to encourage resolution of India's long-standing border dispute with that country, thus helping to reduce India's incentives to acquire the bomb. If steps were successfully taken to encourage efforts by South Korea and North Korea to agree on a framework for peaceful relations, that would lessen the incentives there for acquisition of nuclear weapons.

Because of the complexity of these disputes and the intense emotions involved, the chances for success of such diplomatic initiatives should not be exaggerated. But where conditions appear ripe, possible nonproliferation benefits provide still another reason to try to alleviate or resolve the underlying conflicts.

THE LIMITS OF NONPROLIFERATION POLICY

While checking the growth of incentives for acquiring nuclear weapons is at the heart of efforts to contain the bomb's spread, no one of the preceding initiatives will suffice in itself. Even as potent a policy tool as strengthening U.S. alliances and security ties is applicable to only some of the potential nuclear powers and more often than not needs to be augmented by provision of advanced conventional arms. Rather than seeking one all-encompassing initiative, it is necessary to think in terms of combinations of the preceding measures tailored to reduce specific countries' incentives for acquiring nuclear weapons. It also is important to buttress those initiatives with the threat of sanctions, while taking steps to slow the unavoidable erosion of technical constraints.

Unfortunately, such steps to check proliferation's scope and pace are limited not only by the need for a multifaceted, country-by-country, region-by-region approach but also by the lingering domestic unease within the United States about heightened activism and entanglement abroad and by the fact that nonproliferation policy may be subordinated to other, more pressing foreign policy and national security concerns. Not least of all, there will be times when there may be no compelling initiatives that the United States or other countries can take to reduce a country's incentives to "go nuclear," while available sanctions for inducing restraint may be insufficient. For each of these reasons it is important to begin thinking about responses to the emergence of the next nuclear powers.

6 • BUILDING PROLIFERATION FIREBREAKS

In the next few years, several more countries are likely to carry forward, initiate, or resume nuclear weapons programs. Therefore, in addition to traditional policies to check the bomb's spread, responses must be prepared to build proliferation firebreaks after such occurrences as nuclear tests, withdrawal from or violation of the Treaty on the Nonproliferation of Nuclear Weapons, safeguards violations, gray or black market nuclear transactions, and the many other dramatic events on the road to a full-fledged nuclear force. It will be essential to pursue three partly conflicting objectives—holding down the level of proliferation; checking the regional proliferation multiplier effect; and containing the adverse impact on nonproliferation agreements, norms, and expectations.

Of course, there are dangers in examining now how the United States and other countries might respond to these outcroppings of further proliferation. Perhaps the greatest danger is that focusing attention on such nuclear weapons activity might give rise to the false notion that evermore widespread proliferation is unavoidable, thereby undermining support for nonproliferation efforts. Equally harmful would be the erroneous conclusion that emphasis should be placed on living with, not checking, the spread of nuclear weapons, a conclusion that might be provoked because on some occasions the least bad response will be to come to terms with limited nuclear weapons activity in order to contain it. Even the discussion of the possibility of such accommodation could undercut the threat of sanctions.

But notwithstanding these dangers—and the less creditable, though understandable, preference not to think about what will happen if nonproliferation efforts fall short of their objectives—it is vital to think through this problem now, for additional nuclear weapons activity will present not only a serious threat but also an opportunity. In the crisis atmosphere and aftershock that will follow, it may be considerably easier to obtain public, congressional, and bureaucratic support for new nonproliferation initiatives and to win other countries' acceptance of more effective nonproliferation agreements. Advance planning is necessary, however, to take advantage of that opportunity. Prior analysis also can clarify the benefits and costs of different responses as well as identify ways of reducing those costs. Most important, rules of thumb for responding to increased nuclear weapons activity are needed lest ad hoc responses greatly damage nonproliferation efforts.

RESPONSES TO DRAMATIC PROLIFERATION EVENTS

Covert but Suspected Proliferation

In the early to mid-1980s, several countries—such as Pakistan, India, South Africa, Taiwan, or Iraq—could secretly begin to stockpile nuclear explosive material, manufacture the nonnuclear parts of nuclear weapons, and undertake other preparatory steps for the design and fabrication of nuclear explosive devices. A few of these countries next might covertly acquire a handful of untested, unassembled nuclear weapons, just as Israel is thought to have done. One or more might go a step further and assemble a number of weapons, perhaps even trying to test a bomb without being caught, as South Africa may have done in September 1979.

Western intelligence services are likely to detect possible signs of covert proliferation, and newspaper and television reports based in part on leaked intelligence assessments probably will surface as well. But some, if not considerable, uncertainty is likely to remain about the precise nature of such nuclear weapons activities. Even more highly classified assessments available only to top officials may be no more than best estimates.

There are significant benefits to be gained from containing such proliferation at the covert, if suspected, level. Pressures on neighboring countries to initiate or speed up matching nuclear weapons programs are likely to be less than if nuclear weapons had been openly acquired. The residual ambiguity about the existence of covert nuclear weapons activities could encourage a wait-and-see attitude in neighboring countries, making it easier for those countries' leaders to resist popular and elite demands for a bomb with the argument that there was no need to overreact to what might prove to be a false alarm. Because of its ambiguity, covert proliferation would probably cause less damage than overt proliferation to onlookers' perceptions of their freedom of action. And the reluctance to acknowledge that a bomb program is in progress would reaffirm in a backhanded way the illegitimacy of "going nuclear." Even if some neighboring countries responded with covert activities of their own, this would be less dangerous than a proliferation chain that included overt deployment of possibly technically deficient small nuclear forces vulnerable to accident, theft, and unauthorized use.

But any attempt to convince potential new nuclear powers not to go beyond covert possession of, say, a small number of untested, or even assembled, nuclear weapons entails difficult policy choices. It may require resisting the instinctive inclination to impose sanctions immediately. Sounding the alarm, invoking a full panoply of punitive measures, and seeking to make an example of the suspected country would remove a significant disincentive to overt nuclear weapons activities. For once sanctions are applied, that country easily might conclude that, since little more damage could be done to it, there would be no reason not to proceed. By contrast, the threat of the deferred sanctions would provide a reason for restraint. Further, the imposition of sanctions in response to suspected covert proliferation would intensify public and bureaucratic support in that country for a full-fledged program and increase the like-lihood of its leadership taking emotional, poorly thought-through actions.

Various "carrots" offered by the United States, with the support of other countries insofar as possible, might have a greater chance of inducing the desired restraint. For example, in some instances, sales of advanced conventional arms, diplomatic support, or military assistance probably would dampen incentives for overt nuclear weapons activity. Enhanced U.S. security ties are another, if less widely applicable, means for inducing nuclear restraint. Each of these measures, moreover, would reinforce or create linkages that a country might be very reluctant to jeopardize by an overt nuclear weapons program. And where conditions are ripe, foreign policy initiatives such as renewed efforts to encourage a dialogue between Pakistan and India on their security and nuclear concerns also could help alleviate some of the underlying political pressures for nuclear weapons acquisition.

Coming to terms with covert proliferation in this manner, however, clearly risks encouraging countries to use the threat of "going nuclear" to acquire economic, military, or political benefits. Further, deferral of sanctions in an attempt to hold down the level of proliferation would adversely affect onlookers' perceptions of the risks of covert possession of the bomb. It also could undermine the widespread belief that runaway proliferation is avoidable.

In determining whether to give precedence to holding down the level of proliferation in response to suspected covert nuclear weapons activities, the likelihood that the country under suspicion will stop short of an overt nuclear weapons program is of critical importance. (High confidence intelligence about the country's intentions, therefore, is needed.) Also significant is whether any legally binding nonproliferation obligations have been violated and how much uncertainty exists about the suspected nuclear weapons activities, both of which would greatly affect the ultimate costs of a wait-and-see approach. The likely consequences of overt proliferation in the region—and thus the payoff of holding down the level of activity—also must be assessed. For instance, it might be more justifiable to acquiesce reluctantly to covert Israeli, Libyan, and/or Iraqi nuclear weapons capabilities to avoid the overt nuclearization of such a conflict-prone region as the Middle East; in contrast, the payoffs of avoiding overt South African deployment of nuclear weapons might be less, since the likelihood of neighboring African countries matching such nuclear activities would be slight for some time.

Regardless of which approach is chosen, efforts also will have to be made to reduce the impact of suspected covert nuclear weapons activity on neighboring countries' proliferation incentives. Reaffirming or upgrading existing U.S. security ties where credible and appropriate should help. Although lingering public resistance within the United States to new commitments abroad may preclude formal new security ties, more limited guarantees—such as a presidential pledge of support against nuclear blackmail, perhaps, but not necessarily, with some type of congressional backing—could be useful to tip the balance against matching nuclear weapons activities. But frequently it will be difficult to provide a guarantee that only minimally conflicts with other U.S. regional interests. Particularly in the Middle East, balancing more explicit nuclear security guarantees for Egypt and Saudi Arabia in the wake of Iraqi or Libyan acquisition of untested bombs, with continued U.S. ties to Israel, for example, will be a complicated diplomatic undertaking. Nevertheless, such measures will be necessary to contain the multiplier effect within the region.

The Next Nuclear Tests

Before the middle of this decade, one or more countries may detonate a nuclear explosive device—perhaps claiming interest only in the supposed peaceful uses of nuclear explosives for such engineering jobs as digging canals, harbors, and gas storage domes. The testing of a nuclear device is frequently assumed to represent an irrevocable commitment to the development and deployment of a nuclear force. But there are many intermediate levels of proliferation between a nuclear test and deployment of a full-fledged nuclear force (see Table

139

2). Each is not only a step up the nuclear weapons ladder but also a potential stopping point. Further, some countries may not have compelling incentives to go beyond an initial nuclear test. For example, if Pakistan eventually matches India's 1974 PNE, this alone might suffice to assuage Pakistani national pride, provide sought-after prestige in the Moslem world, bolster Zia's domestic political position, and lessen concern about a strategic gap with India. As well, fear of disrupting Pakistan's new strategic relationship with the United States and of losing a nuclear arms race with India could be a potent disincentive to additional Pakistani nuclear weapons activity.

Significant payoffs would result if the overt nuclear weapons activities of new nuclear powers could be held to the lowest levels of proliferation. The proliferation multiplier effect within a region as well as the corrosive impact on global nonproliferation norms and expectations is likely to be less severe. Also

Table 2. Some Levels of Initial Nuclear Weapons Capability

- Detonation of a PNE
- Testing of a single nuclear weapon
- Additional nuclear tests to transform the initial device into a usable weapon
- Central stockpiling of first-generation fission weapons with ad hoc plans for using available aircraft for delivery
- Detailed planning and preparations to determine requirements of deploying a nuclear force
- Creation on paper of a separate nuclear strike force
- Training of selected air force, ground, or naval units for nuclear operations—without nuclear weapons
- A series of nuclear tests to develop well-packaged, reliable, and easily deliverable fission weapons
- Institution of rudimentary warning procedures and command-and-control arrangements—but still without dispersal of nuclear weapons around the country
- Deployment of nuclear weapons from central storage to limited numbers of military units and bases
- Availability of a strategic nuclear force targeted on potential enemies with its own personnel, budget, and standard operating procedures; periodic exercises; and well-defined doctrine

the problems of mitigating proliferation's adverse consequences would be eased if these countries could be induced not
to move up the nuclear weapons ladder.

Again, as in the case of covert but suspected proliferation,
the immediate imposition of a broad range of sanctions following the next nuclear test probably would hinder any attempt to
contain the level of proliferation by removing one incentive for
nuclear restraint. It could be necessary, moreover, to waive
congressionally mandated restrictions on military and economic assistance, as that assistance, in conjunction with new or
reaffirmed U.S. security guarantees, might be required to reduce the security-related incentives that may have led the
country to "go nuclear" in the first place. Diplomatic initiatives
to convince the new nuclear power that its security actually
could be reduced by heightened nuclear weapons activity also
might be appropriate.

But such a tempered response to induce restraint after a nuclear test would incur considerable costs. Selling advanced conventional arms to Pakistan and India, for example, might
lessen the chances of a South Asian nuclear arms race but fuel
a conventional one. Reaffirming U.S. security ties to Pakistan
would mean accepting the risks and burdens of foreign policy
activism as well as the possible costs of deteriorating relations
with India. Still other countries, seeing the United States providing sweeteners for nuclear restraint even after so dramatic
an act as a nuclear test, might threaten to "go nuclear" to win
comparable benefits. But the highest cost of coming to terms
with such limited proliferation in an effort to avoid even more
extensive nuclear weapons activities is that the failure to make
an example of the next country to test a nuclear device would
weaken considerably other countries' perceptions of the risks
of testing a nuclear weapon and of the likelihood of runaway
proliferation. Nonetheless, if no legally binding obligation has
been violated and there is a good chance of inducing the new
nuclear power to stop even after a nuclear test, the least bad
alternative once again may be to temper the U.S. response.

Furthermore, the adverse impact on the deterrence of still
other countries' nuclear weapons activities of such a decision

not to impose a full panopoly of sanctions might be lessened somewhat if the United States at least cut off nuclear exports to the country that had detonated a nuclear explosive device. While an automatic cutoff of nuclear assistance would undermine to some degree U.S. efforts to provide incentives for nuclear restraint, this would be a price worth paying in order to keep other countries from concluding that there are few costs to comparable, if not more far-reaching, nuclear weapons activities. By that cutoff—which would be most effective in concert with cutoffs by other countries—the United States would both demonstrate its readiness to impose frequently painful nuclear energy-related sanctions and implicitly threaten an even stronger punitive response where mitigating circumstances were lacking. The costs of occasional accommodation also might be lessened by statements by the President, other officials in the executive branch, and Congress stressing that the deferral of other sanctions and the waiving of congressionally mandated restrictions were the exception rather than the rule and that such a tempered response was warranted for only as long as the new nuclear power exercised restraint.

Coinciding with such an attempt to contain the level of proliferation even after a nuclear test, measures also need to be adopted to check the proliferation multiplier effect within the region. Military and/or economic assistance, arms sales, renewed or even new security ties, diplomatic support and persuasion, and the threat of sanctions all may be useful in prevailing on neighboring countries not to acquire nuclear weapons. A mixture of initiatives is likely to be needed, and its detailed components will vary from case to case.

While the primary objective following the next nuclear test will be to contain its adverse repercussions, there also will be an opportunity to launch new nonproliferation initiatives. Even though the specifics of a new policy will depend on the situation, some directions it might take are already evident. For example, capitalizing on the probable lessening of public and congressional reluctance to take on new commitments abroad, strengthened or even new security guarantees might be offered to some countries not only in the immediate vicinity of the new nuclear power but also in other proliferation hot-spots. Or, the time might be ripe to push for international acceptance of tightened controls over gray area exports not yet covered by existing restrictions—particularly if these exports played a role in the test. Steps also could be taken to establish new procedures and institutions governing the global use of nuclear energy. Fresh diplomatic initiatives, running the gamut from steps to alleviate the political disputes that could fuel proliferation to support for nuclear free zones, might lessen the chances of additional covert or overt nuclearization in conflict-prone regions.

Clear Violations of Legally Binding Nonproliferation Obligations

Neither covert nuclear weapons activities nor detonation of a nuclear explosive device would necessarily entail clear-cut violation of a legally binding nonproliferation obligation. After all, many of the countries that might "go nuclear" are not parties to the NPT and thus are not bound by its provisions proscribing manufacture or acquisition of nuclear weapons or other nuclear explosive devices. Those countries that are parties to the NPT can exercise their right of withdrawal, citing grave threats to their security. Reliance on indigenous facilities to produce nuclear explosive material also would avoid a clearcut safeguards agreement violation. And though American bilateral agreements for nuclear cooperation, as well as those of some other nuclear suppliers, contain clauses banning the use of transferred facilities, components, or materials for the manufacture of nuclear explosives, such use can be avoided.

It is quite likely, however, that at least one country will violate a legally binding nonproliferation obligation on its march to the bomb in the decade to come. Lack of a firm, nonaccommodating response to such a violation—in contrast to covert but suspected proliferation or even a next test—would have extremely serious costs for proliferation policy.

Failure to oppose a clear-cut violation would markedly lessen many countries' perceptions of the risks of acquiring nuclear weapons. For example, if covert or overt acquisition of nuclear weapons by a party to the NPT were not met with a strong response, other parties to the Treaty would be more willing to violate its provisions. Moreover, the Treaty itself would lose credibility. Failure to respond forcefully to a clear violation of a bilateral agreement for nuclear cooperation also would undermine the readiness of the major nuclear suppliers to abide by existing nuclear exports constraints and to accept new ones; continuing economic pressures, a fear that proliferation was becoming uncontrollable, and the fact of having once capitulated all would weaken their resolve to stand firm. An unpunished safeguards violation also would demoralize the International Atomic Energy Agency, making its inspectors and officials less ready to pursue intrusive investigations; to raise questions about ambiguous activities; to insist upon the widest interpretation of their rights of access to all areas of nuclear facilities; and, at higher levels within the IAEA Directorate of Safeguards, to push its side of disputes. Consequently, confidence in the ability of safeguards to detect illegal diversion of nuclear explosive material would be undermined and in regions where traditional rivals are suspicious of each other's long-run nuclear intentions, the result is likely to be heightened pressures to be the first to "go nuclear."

Not only are these costs of inaction considerable, but the chances of holding down the level of proliferation are slim. The very act of violating a legally binding nonproliferation obligation represents a seriousness of commitment to acquire nuclear weapons that belies the prospects for inducing restraint. Consequently, limiting the damage done to global nonproliferation norms, agreements, and expectations by imposing a full panoply of sanctions probably should take precedence. And even if other vital foreign policy or security considerations dictated making an exception in any one case, it still would be advisable at least to cut off nuclear exports to that country in an attempt to reduce the damaging repercussions of such a violation.

The particular package of sanctions to be applied in addition to an automatic nuclear exports cutoff will vary from country to country. Their purpose in all cases, however, would be to inflict significant costs on the target. Because sanctions' impact would be increased considerably should U.S. efforts be made in concert with those of several other advanced industrial countries, steps ought to be taken now to encourage the growing international recognition of the importance of standing behind legally binding nonproliferation obligations.[1] And while the United States and other advanced industrial countries still may be reluctant to carry out sanctions at the expense of other policy objectives, it is likely to be easier to win support for a strong multilateral response following a violation of nonproliferation obligations than after any other dramatic proliferation event.

In the event of a clear-cut safeguards violation, the United States also might take advantage of the shock to push for international agreement on and implementation of improvements in the IAEA safeguards system. Specific measures to be pursued might include expanded provision for unannounced IAEA inspection, unrestricted IAEA access to all parts of safeguarded facilities, more rapid and wider disclosure of IAEA findings, and the right of special inspections by the IAEA or any appropriate regional agency at the request of countries with reason to question a neighbor's nuclear activities. Measures less closely related to the particular violation—such as agreement among the major suppliers to cut off nuclear exports automatically in response to violations of any legally binding nonproliferation obligations, steps to tighten controls on gray area exports, and cooperation in establishing new institutions to govern the global use of nuclear energy—could be sought as well.

Withdrawals from the NPT

There has been growing discontent with the NPT. Many developing countries resent the nuclear powers' reluctance to take steps to slow the nuclear arms race and to assist other countries in acquiring peaceful nuclear technology as the Treaty requires.[2] Thus, one or more nuclear have-nots may yet exercise their right to withdraw from the Treaty. Under Article X of the NPT:

> Each party shall in exercising its national sovereignty have the right to withdraw from the Treaty if it decides that extraordinary events, related to the subject matter of this Treaty, have jeopardized the supreme interests of its country. It shall give notice of such withdrawal to all other Parties to the Treaty and to the United Nations Security Council three months in advance. Such notice shall include a statement of the extraordinary events it regards as having jeopardized its supreme interests.

Should a country withdraw, steps would have to be taken to limit the damage done to the Treaty. While avoiding statements exaggerating the extent of the crisis that could only heighten concern about a breakdown of the NPT, the United States should make sure that the provisions of Article X—particularly the requirement that the party justify its withdrawal—are strictly followed. Action by the U.N. Security Council might be sought as well in support of the contention that the right of withdrawal from the NPT is not to be lightly exercised. Complacent acquiescence to a withdrawal would further weaken the norm of nonproliferation and make it more likely that other countries uneasy with the strictures of the NPT but formerly unwilling to pay the political price of withdrawal might now seek to do so.

Measures to limit the impact of a withdrawal from the NPT on neighboring countries also will be needed. Holding down the level of nuclear weapons development by the withdrawing party, security reassurances to the neighboring countries themselves, quiet diplomacy to urge restraint, and the threat of sanctions all could be useful in preventing a regional multiplier effect.

It may not be advisable, however, to make use of a withdrawal from the NPT as a catalyst for reform of the Treaty itself. All sorts of changes are likely to be proposed in such a crisis atmosphere—from efforts to provide preferential access to nuclear exports for NPT parties to explicit recognition of the right of nonnuclear weapons states to use plutonium as a fuel—not all of them beneficial. Moreover, a long, contentious international debate about those reforms, reiteration of the

Treaty's defects, and criticism of the nuclear haves by the nuclear have-nots for not living up to their treaty obligations
could easily weaken the Treaty's legitimacy and exacerbate that
already simmering resentment of the Treaty among quite a
few developing countries. Still more withdrawals and the un-
raveling of the NPT could be the unintended result.

Nuclear Gray and Black Marketing

It is quite possible, if not likely, that there will be gray market
nuclear purchases, government-to-government nuclear weapons cooperation, hiring of nuclear mercenaries, and even theft
of nuclear explosive material in the years to come. Following
any one of these occurrences, it clearly would be appropriate
to reassess existing U.S. measures for handling these threats
and to devise new policies to reinforce them. For example, the
nuclear exports controls of the United States and, if possible,
those of other suppliers might be extended to include newly
identified gray area components and materials; procedures for
monitoring and pooling intelligence about attempts to circumvent exports controls or hire nuclear mercenaries could be
strengthened to provide earlier warning of gray market purchases and permit multilateral preventive action; and, in addition to taking steps to recover the stolen material, protection
and security measures against black market theft might use-
fully be reevaluated and any gaps closed.

Particularly threatening to U.S. nonproliferation efforts
would be government-to-government nuclear weapons cooperation, including even the transfer of nuclear weapons design
information and nuclear explosive material by a new nuclear
power to a nonnuclear country. Once countenanced, such a
transfer would set a precedent for the outright buying and selling of nuclear explosive material, designs, and weapons themselves; undermine the norm of nonproliferation and increase
perceptions of runaway nuclear weapons activity; and augment
the likelihood that irresponsible governments and radical subnational groups will gain access to nuclear weapons.

Confronted by intelligence warning of such cooperation between governments, the United States and its allies should

stress through every possible diplomatic channel that they view such nuclear weapons activity—especially if it involves the actual transfer of nuclear explosive material and design data—to be beyond the boundaries of tolerable international behavior. To drive that point home, any available economic and political leverage ought to be brought to bear. If that fails to derail the nuclear transaction, other measures—none of which admittedly is very attractive—would have to be considered.

Direct military action by the United States, using either conventional forces or covert special-action teams, could be employed to destroy the facility producing the soon-to-be-transferred nuclear explosive material or to intercept or destroy that material in transit. But this presupposes detailed intelligence about the location, vulnerabilities, and defenses of the nuclear facilities and/or about the timing and procedures for the transfer that may not be available. Besides, direct military action would be ineffective against transfer of design information or technical assistance, barring willingness to assassinate foreign nuclear scientists and technicians. In addition, the foreign policy costs of direct military action—not the least of which is the risk of confrontation with the Soviet Union—could be unacceptably high.

Regardless of U.S. caution, a threatened neighbor of the proposed recipient could seek to intercept or destroy the material or weapon as it is being transferred. For example, Israel attempted to scare off German scientists at work on a rocket program for Egypt in the early 1960s; is allegedly responsible for the assassination of Iraq's leading nuclear scientist, Yihya Al-Meshad, in 1980;[3] and, in 1981, destroyed an Iraqi nuclear reactor it believed was to be used to make bombs. Should U.S. intelligence uncover plans for such direct action by a neighboring country, a decision would have to be made as to whether to forestall, turn a blind eye, or even indirectly support it (although this last possibility would not only risk a clash with the Soviets but quite possibly also would be in tension with U.S. foreign policy interests in the region).

As unattractive as military action might be, abstinence—particularly should access to nuclear weapons by unstable and irresponsible governments be at issue—could prove too costly. In such a situation, direct or indirect military action might be the least bad alternative. Still, reliance on diplomatic means and the threat of sanctions to deter or abort a government-to-government transfer of nuclear weapons design information, components, and nuclear explosive material, or even a bomb itself, clearly are preferable.

RULES OF THUMB

Taken together, the rules of thumb contained within the preceding discussion provide a framework for responses to the outcroppings of additional proliferation in the years ahead. But it is important to remember that these suggested guidelines are only part of a broader strategy for containing proliferation whose bedrock is measures to slow the pace and limit the scope of such nuclear weapons activities. Only that larger perspective can legitimize coming to terms occasionally with limited nuclear weapons activities in the interest of holding down the level of proliferation and can head off demoralization in the wake of the next dramatic proliferation events.

7 · MITIGATING PROLIFERATION'S CONSEQUENCES

Despite efforts to check the bomb's spread and to build proliferation firebreaks, it probably will not be possible to head off the deployment of at least rudimentary nuclear forces in one if not more conflict-prone regions during this decade. As a result, the first decades' nuclear peace may be shattered. More systematic thought must be given, therefore, to measures for mitigating proliferation's regional and global consequences. While not without risk, such contingency planning—as with the delineation of rules of thumb for responding to dramatic proliferation events—can considerably enhance the effectiveness of later policy initiatives. And an appreciation of the limits of even those initiatives makes clear yet again the fundamental significance of traditional nonproliferation policies.

FOSTERING STABILITY IN NEWLY NUCLEARIZED REGIONS

One place to start is with measures to lessen the danger of local nuclear conflict, to bolster the security of nonnuclear allies and friends threatened by countries with growing nuclear weapons capabilities, and to foster more stable structures of regional security. These proposed steps are consistent with the continued, if more cautious, pursuit of U.S. interests in these newly nuclearized regions and would not require a major change of course for the United States.

Technical Assistance to New Nuclear Powers

Once a country is openly committed to moving up the nuclear weapons ladder, it might seem eminently reasonable for the United States to assist that new nuclear power to develop a safer, more tightly controlled, and less accident-prone nuclear force. Such assistance would lessen the chances of an accidental or unintended nuclear exchange in these regions, reduce the likelihood of nuclear coups d'etat that could spill over into wider regional conflict, and perhaps prevent terrorist theft of nuclear weapons.

But providing technical assistance would entail significant costs. Until now, nuclear weapons have been seen as difficult to control and therefore dangerous to their owners as well as to other countries. Particularly for the many politically unstable countries that could "go nuclear," this concern over loss of control is a significant disincentive to both the initial acquisition of the bomb and to movement to the higher levels of nuclear weapons activity. If assistance from the United States on nuclear weapons safety and security were forthcoming, this disincentive would erode. Moreover, such assistance probably would reveal information about the design and engineering of more efficient nuclear weapons; decrease fear of unauthorized access and use, thereby removing a major constraint on large-scale nuclear weapons production; facilitate these weapons' military deployment; and enhance the operational capabilities of the ensuing nuclear force. As a result, the military effectiveness of a new nuclear force would increase, which could lead to a more belligerent diplomatic posture, less readiness to compromise in a crisis, and possibly even a greater willingness to use nuclear force. By providing technical assistance, the United States also in all probability would become more closely linked with the recipient, increasing the likelihood of military entanglement in a regional conflict.

To a certain extent, these costly repercussions may be alleviated if such assistance is given at the proper time. If the United States were to indicate a readiness to help ensure the safety and security of any new nuclear forces immediately following the next test of a nuclear device, this would greatly reduce the disincentives of many countries to acquire a nuclear weapons capability. By contrast, waiting until the onset of runaway proliferation would be too late. In between these two extremes, however, there probably will be some point—perhaps the deployment of nuclear forces in a particularly conflict-prone region such as the Middle East, a failure to hold the line at isolated pockets of proliferation, or a nuclear theft—at which the benefits of providing U.S. assistance would outweigh the adverse effects. Even then, it would be best to maintain a low profile, making assistance available on an informal, ad hoc basis, so as to reduce regional and global perceptions of the lessened risk of acquiring the bomb.

Should it become U.S. policy to provide technical assistance to certain new nuclear powers, decisions on how much and what kind of assistance, and to whom, still would have to be made. All countries might be given assistance in designing and implementing an overall physical security system to protect nuclear weapons storage sites.[1] Valuable information and procedures derived from more than thirty years of American experience with these weapons and covering matters ranging from how to set up perimeter fencing and lighting for a nuclear weapons site to the selection of emotionally stable, reliable personnel—along with appropriate materials and components—could be offered. Basic command-and-control concepts—such as the "two-man rule," which requires that two individuals, each acting independently, be present at every level of operation or handling of nuclear weapons—also could be discussed with any interested new nuclear powers.[2]

In contrast, the sale or transfer of advanced command-and-control devices—such as permissive action link (PAL) systems—designed to impede unauthorized access electronically could considerably augment the military versatility of the recipient's nuclear force and might be better reserved only for long-standing allies.* And information on accident-proofing nuclear

---

*For example, with such advanced technology, training exercises and practice alerts could be carried out with actual weapons, battlefield deployment of nuclear weapons and more dispersed basing would become more feasible, and potentially useful information about nuclear weapons design would be imparted.

weapons also might best be restricted to long-standing allies because discussion of the more sophisticated techniques would reveal weapons design concepts. Restricting the types of assistance provided, however, would to some extent reduce the potential payoff of the overall attempt to enhance the safety and security of these new nuclear forces. Thus, in some situations, there still could be considerable risk of accidental or unauthorized nuclear use.

"Nuclear Learning"

The United States also might seek to influence the strategic thinking and practices of countries committed to building up their nuclear arsenals in an effort to lessen the chances of a local nuclear conflagration. Formal and informal contacts could be pursued to encourage the perception that, because of their awesome destructiveness as well as the danger that even the limited use of nuclear weapons will escalate, these weapons are not simply more advanced conventional weapons to be used when military efficiency dictates but are primarily instruments of deterrence. Uncertainties about the performance of even sophisticated nuclear forces, possible failures of command and communication, the difficulties of ending a nuclear war, and a myriad of other complications also could be stressed to discourage adoption of a nuclear warfighting posture.

Steps also might be taken to make it less likely that new nuclear powers would adopt a virtually all-nuclear defense posture—one in which conventional forces are only the thinnest of tripwires to all-out nuclear conflict. At a minimum, the United States could urge such countries to maintain robust conventional forces, warning of the dangers of overreliance on nuclear weapons. Future requests from new nuclear powers with long-standing ties to the United States—a group that could include Israel, South Korea, and, to a lesser degree, Pakistan—to purchase advanced conventional arms might be approved as well, lest they be forced to rely solely on nuclear weapons. Of course, U.S. arms sales to new nuclear powers would undermine the threat of sanctions elsewhere, while heightening the risk of becoming involved in a regional—and possibly nuclear—conflict. There is, however, no escape from this dilemma.

The United States also should try to convince these countries not to engage in practices—such as planning only for unrestrained nuclear use against cities—that would result in the maximum number of fatalities should deterrence fail and war break out. Both humanitarian sensibilities and narrower U.S. interests suggest trying to reduce the destructiveness of such a small power nuclear war. Moreover, even if it rapidly breaks down, initial restraint in the use of nuclear weapons might still provide one last chance for the opposing countries to call a halt or for outside forces to intercede before resort to the destruction of cities begins. Admittedly, it might be next to impossible to convince a weak country seeking to deter a stronger rival to renounce in advance the threat of city-busting attacks. And, at least initially, many new nuclear powers may only possess crude terror weapons. Or, despite agreement in principle, insufficiencies of command, control, and communication capabilities may preclude the adoption of even limited restraint and flexibility. Nonetheless, over time, more countries are likely to become receptive to such arguments against unrestrained use of nuclear force and technically better able to exercise control.

Much of the "nuclear learning" that the United States might seek to convey probably would be uncovered sooner or later by a new nuclear power. But U.S. efforts would speed that process—compensating for the more limited time that new nuclear powers in conflict-prone regions will have to adjust to living with nuclear weapons. They also would help to avoid dangerous missteps.

Support for Nonnuclear Countries

A readiness to support nonnuclear allies and friends that find themselves threatened by new nuclear powers would contribute as well to the stability of such nuclearized regions. Earlier pledges of support against nuclear blackmail or attack might be reaffirmed by well-publicized, high-level diplomatic meetings. For countries in which vital U.S. interests are at stake, prior pledges might be upgraded and formalized as executive agreements or treaties. New pledges of support might even be offered. And, to heighten the credibility of old as well as new promises of support, it could be announced that some U.S. strategic forces—say a limited number of ICBMs—were being readied to retaliate on behalf of any one of these countries.[3]

The United States also might consider indirect steps to buttress the defenses of its allies and friends against nuclear attack. In most cases, the foreseeable threat to these countries in the 1980s is likely to be an attack by a small nuclear force that relies on aircraft for delivery. Thus, supply of and training in the use of high performance aircraft, surface-to-air missiles, and associated air defense warning and battle management systems could have a high payoff for protecting the most critical targets. And rapid advances in ballistic missile defense technology hold out the prospect of highly effective defense against the limited missile capabilities that a few new nuclear powers might be able to acquire. But while the United States could provide the personnel needed to operate and maintain such a missile defense system for threatened countries, there will be few situations—one might be limiting damage to Saudi oil facilities—warranting the risks of direct U.S. entanglement in these conflicts.

Sales of advanced conventional arms also could enhance the security of nonnuclear allies and friends in these regions. Bolstering their conventional defenses might both help deter conventional conflicts that could escalate to a nuclear confrontation and permit the recipient to handle limited incidents on its own, without inviting direct U.S. military intervention. These sales also would show the vitality of U.S. security commitments, thereby strengthening the recipient's willingness to resist nuclear threats as well as lessening the chances that the new nuclear power will miscalculate the readiness of the United States to come to the assistance of its nonnuclear ally and launch an attack.

Support for a long-standing ally might even include the deployment of American ground forces and naval units. Their mission could range from enhancing deterrence and providing reassurance by showing the flag to actual military operations against an invading nuclear-armed neighbor. But there are probably few instances in which sufficiently vital U.S. interests would be at stake to justify the heightened risks of military intervention in a newly nuclearized region.

While U.S. support for nonnuclear allies and close friends occasionally will be necessary to contain the adverse regional consequences of proliferation, the scope and instances of that support are likely to be less than such abstract analysis might recommend. Some segments of the American public will oppose taking the greater risks involved. Moreover, in some regions—such as the Middle East—the diplomatic and domestic maneuvering necessary to provide security support to one country—for instance, Egypt—without alienating another— say, Israel—might be so complex as to forestall action. But to the extent that such external support for nonnuclear countries is not forthcoming, the danger of regional nuclear blackmail and conflict will remain unalleviated.

Managing Old Alliance Ties with New Nuclear Powers

Several of the countries that could become nuclear powers in the next decades—Israel, Pakistan, South Korea, Taiwan, or even Japan, for example—currently have political and security ties to the United States. Even in cases in which an erosion of those ties would have helped precipitate nuclear weapons activities, a residual security relationship may remain, posing a dilemma for the United States.

Faced with an ally's deployment of nuclear weapons, there would be a strong presumption that the United States should follow the previously delineated rules of thumb and sever those alliance ties as part of a multifaceted punitive response intended to stand behind the norm of nonproliferation and to influence other countries' perceptions of the costs of acquiring nuclear weapons. The greater risks of continued involvement with an ally that independently controls nuclear weapons would reinforce the arguments for that course of action. Further, if the ally's nuclear force were vulnerable to an opponent's surprise attack or subject to some other technical deficiency that could not be easily remedied by outside assistance, the heightened chances of local nuclear conflict would bolster the case for decoupling.

But preserving the alliance relationship might better serve
the U.S. objective of fostering security and deterrence in the
region. Once the linkage has been severed, the United States
would lose its remaining leverage and opportunities for influencing the nuclear doctrine and posture as well as the regional
behavior of the new nuclear power. Punitive decoupling would
shake regional stability even further at a time when there is
high tension and uncertainty and when neighboring countries
are reassessing their own need to acquire nuclear weapons. Besides, other U.S. economic and security interests in the region,
as well as domestic American politics, also might argue against
severing alliance ties.

The status of the alliance connection itself will be significant
in determining whether the United States should sever or
maintain its security relationship. If an ally's decision to acquire nuclear weapons is not triggered by the prior erosion of
the U.S. guarantee—as would be so, for example, if Israel
openly deployed nuclear weapons in response to Iraqi nuclear
weapons activities or if, despite ties to the United States, Japan
overreacted to a South Korean bomb by "going nuclear"—U.S.
interests might be better advanced by maintaining the original
ties. And, by using diplomacy to express its displeasure rather
than rushing to embrace the new nuclear power, the United
States also could lessen somewhat the adverse nonproliferation
impact of such accommodation.

Steps also could be taken to reduce the risks of maintaining
ties on occasion with a newly nuclearized ally. For example, the
United States might initiate discussions with that ally to clarify
respective positions on the use of nuclear weapons, thus both
reducing the unexpectedness of any such recourse and facili-
tating preparations for an appropriate U.S. reaction. Those exchanges also might provide opportunities for the United States
to influence its ally's nuclear doctrine and posture. In addition,
the establishment of formal mechanisms to coordinate defense
planning and policies, which would help avoid actions at cross-
purposes in a crisis or conflict, could be broached. Still, even
with these measures, successfully managing an alliance rela-

tionship with an ally that has begun to deploy nuclear weapons will be a complicated and difficult task.

Regional Nuclear Arms Control

In newly nuclearized regions—such as South Asia or the Middle East—the United States could encourage a variety of arms control measures that would reduce the threat of local nuclear conflict and enhance the security of its allies and friends.[4] A good place to start would be confidence-building measures to lessen reciprocal fears of surprise attack and to reduce the chances for nuclear accidents or misunderstandings. For example, where traditional rivals each possess bomber forces that are vulnerable to surprise attack, the United States could encourage agreements that permit each side to station observers in the other country and set up operating restrictions designed to prevent incidents that could be mistaken for the first signs of a surprise nuclear attack.[5] In addition to diplomatic support, the United States could provide technical assistance for the establishment of hot lines between countries such as India and Pakistan or even Israel and Iraq. The availability of a rapid and secure channel of communication could reinforce other efforts to keep a low-level crisis from escalating; bring a limited conventional clash to a close before it erupted into a nuclear conflict; or offer a last chance to avoid a nuclear war following, say, a nuclear weapons accident or unauthorized strike.

If and when traditional rivals begin to deploy nuclear weapons, quantitative and qualitative limitations on their forces should be encouraged by the United States. In a region such as Latin America, where uncertainty about a rival's intentions may have triggered decisions to acquire the bomb, the newly nuclearized rivals might be reluctant to move up the nuclear weapons ladder but fear that a failure to do so would give their rival a political and military advantage. Still elsewhere, two countries, perhaps Israel and Egypt, might have an incentive to restrain competition between themselves even while engaging in a nuclear arms race with a third country such as Libya or Iraq. Under these conditions, U.S.-supported limitations on

the size, characteristics, deployment, operation, or modernization of each side's nuclear forces could check all-out arms racing. The United States might not only delineate possible
restraints but also more actively assist in verifying any accords—
just as U.S. forces now verify the Sinai disengagement agreement between Israel and Egypt.

A local agreement not to use nuclear weapons first also
would be a significant support to regional stability. Though
such an agreement might be broken in time of crisis, under
normal conditions it would somewhat allay the fear of surprise
attack that could heighten the danger of unintended war and
encourage arms racing. It also would reassure nonnuclear
neighbors. Besides, breaking a no-first-use agreement might
prove more difficult than sometimes thought because such an
agreement would lead countries to reorient their military training, exercises, and capabilities. In addition, its more subtle influence on thinking about and planning for conflict would be
beneficial in itself, moving both sides away from mutually
threatening postures designed to get in the first blow.

Some new nuclear powers might welcome regional no-first-
use pledges. For instance, fears of destructive attacks against
each other's cities could make both India and Pakistan receptive to a no-first-use agreement, particularly if Pakistan concludes that using nuclear weapons first would not provide battlefield advantage. Such an agreement between Egypt and
Israel might signal their intentions to avoid the use of nuclear
weapons against each other even though both may have to deploy those weapons in response to the nuclear weapons activities of other countries in the region. And even though the impact of U.S. support for regional no-first-use agreements
would be weakened by American reluctance to accept that principle for itself in Europe,* encouragement of locally originated
initiatives still would be of benefit in promoting their acceptance.

The United States also ought not overlook measures for reducing the risk of minor incidents and larger clashes involving

---

*That reluctance, to recall, is based on concern about the impact of Ameri-
can acceptance of no-first-use on NATO defense and West German nuclear
weapons incentives.

conventional forces that could trigger a nuclear conflict. Possible measures it could foster include: the creation of demilitarized zones; prior notification of, and limits on the size and duration of, military exercises; arrangements for neutral truce observers; acceptance of mutal surveillance of border areas, possibly with remotely controlled mechanisms to provide early warning of attack such as now monitor movement in the Mitla Pass in the Sinai; and information exchanges on troop dispositions.[6]

Nevertheless, there are limits to what regional arms control measures—or for that matter, this set of policies for fostering regional stability and deterrence—can accomplish. Unless the underlying political disputes and conflicts of interest can be alleviated by other means, such measures will be only partly successful. The danger of local nuclear conflict will be lessened but not ended.

CONTAINING THE GLOBAL REPERCUSSIONS

Both technical assistance to new nuclear powers and regional arms control initiatives will contribute indirectly to containing the global repercussions of the spread of nuclear weapons to conflict-prone regions. Assistance in developing an adequate security system for nuclear weapons, for example, would hinder nuclear theft—helping to close one route that a subnational terrorist group might take to obtain a bomb to threaten the continental United States or an American ally in Europe. It also would lessen the chances of unauthorized use of nuclear weapons in local conflict, thereby dampening one flash point of Soviet-American confrontation. Regional arms control agreements designed to reduce the risk of small-power nuclear conflict would have a comparable effect. Other steps specifically conceived to ensure the ability of the United States to protect its vital interests in newly nuclearized regions, to lessen the danger of a U.S.-Soviet confrontation, to defend the American homeland from small-power or terrorist nuclear threats, and to contain the domestic political repercussions of proliferation must be pursued as well.

Military Intervention in Newly Nuclearized Regions

The United States is developing an enhanced capability—the newly created Rapid Deployment Force (RDF)—to project military power into the Third World in order to protect U.S. economic and security interests. While only the most vital interests would justify military intervention where nuclear weapons are controlled by hostile small powers, such situations may arise. For example, within a newly nuclear Middle East and Persian Gulf, it might be necessary to provide support for a nonnuclear Saudi Arabia threatened by a nuclear-armed Iraq or to take measures to protect the oil fields from nuclear attack. But the control of even crude nuclear weapons by a regional opponent will complicate and constrain such projection of U.S. military power.

Since even only a small number of relatively crude nuclear weapons would pose a serious threat to U.S. airborne, naval, and other forces, it will be necessary to decide whether or not to strike that nuclear weapons capability at the outset of intervention. While there is likely to be considerable pressure from military and some civilian policymakers for a preemptive strike, other considerations argue against it.

In the absence of timely and precise intelligence about the location, size, and vulnerabilities of the opponent's nuclear weapons, sufficiently high confidence of destroying its nuclear weapons stockpiles by a nonnuclear first strike using tactical or long-range aircraft might not be attainable. A commando raid could fail for the same reasons. And, if only a few nuclear weapons survived, they could inflict great damage on U.S. intervention forces. Though reliance on very low-yield nuclear weapons to destroy that nuclear threat would avoid these uncertainties, the heavy onus of again using nuclear weapons—possibly for the first time since Nagasaki—would fall on the United States. Regardless of the putative threat, American judgment would be questioned and criticized by many allies, while U.S. relations with developing countries would be negatively affected. By breaking the nuclear taboo, the United States also would make it easier for another country to use its nuclear weapons and for proponents of acquiring the bomb to argue their case. Thus, in light of the intelligence uncertainties and political costs, it may be advisable instead to deter a small power from using nuclear weapons against American intervention forces by threatening that the retaliation for such use would be an attack not only on that new nuclear power's military capabilities but also on its industrial base and other valuable economic assets.

But deterrence might fail—whether because of unauthorized or accidental use of nuclear weapons; because a desperate leader, thinking all was lost, might strike out at the United States; or because of disbelief that the United States would carry out its threat. Consequently, adaptations in the tactics of intervention that would limit the impact of the use of nuclear weapons against U.S. forces by a small nuclear power should be incorporated into planning for the RDF.[7] For example, in contrast to defense planning in the Vietnam War, when American aircraft carriers were routinely stationed close off North Vietnam in the Gulf of Tonkin, naval units, possibly operating in the Indian Ocean, could be directed to provide air support for ground forces from a farther and safer distance. Greater dispersal of landing forces might be adopted to reduce the value of individual targets and the chances for disrupting the overall operation by the use of only one or a few nuclear weapons. In both the training of individuals and in unit exercises of the RDF, more effort ought to be placed on preparations for operating in an environment in which a nuclear weapon might be or had been used by a small nuclear power. Any necessary changes in force structure and equipment for such "nuclear-scared" operations—from stress on highly mobile and small self-contained units to air filters for armored vehicles and personnel carriers that allow safe passage through fallout areas—should be identified and implemented. Further, the capability must be developed to provide timely and accurate tactical intelligence about the disposition of hostile nuclear weapons in order to permit the military commanders on the spot to take protective measures on warning of possible use or, if unavoidable, to allow the President to order a preemptive strike against those weapons.

None of these adaptations entails major changes in the tactics, training, structure, or equipment of U.S. intervention forces. Nevertheless, there is considerable reluctance to think about them. That reluctance stems in part from a reasonable concern that even limited contingency planning would erode nonproliferation efforts. But, to a greater degree, it is the result of the belief that the problems a small nuclear power might pose for U.S. intervention forces will be solved automatically through U.S. efforts to deal with the Soviet threat. However, while little new equipment may have to be procured to handle the small-power nuclear threat, it is wrong to conclude that there would be no unique tactical, training, intelligence, and other requirements for dealing with that lesser threat. Failure to take seriously the need to identify and meet these requirements will gravely hinder efforts to protect U.S. interests in these newly nuclearized conflict-prone regions. And even though the adaptations required are changes at the margin, they are far from marginal changes.

Mitigating the Increased Risk of Superpower Confrontation

For reasons ranging from the greater number of flash points to the heightened tempo of events, the deployment of small nuclear forces by countries in conflict-prone regions is likely to enhance the risk of Soviet-American confrontation. As an initial step to reducing that risk, the United States should initiate high-level discussions with Soviet officials to clarify how each country defines its regional interests, what courses of action and options each considers legitimate, which Soviet and American moves would be unacceptable to the other superpower, and how each thinks the presence of locally controlled nuclear weapons would affect both sides. A more accurate understanding of the other side's thinking could help to avoid miscalculation and unintended confrontation. And, where views conflict, to the extent possible, compromises with the Soviet Union—or at least agreement to disagree—should be sought.

A Soviet and American pledge to give advance notice, if not limited prior consultation, before providing military assistance, let alone taking direct military action, in a newly nuclearized region also would reduce the risk of superpower confrontation and ought to be pursued. Commitment to the tacit rule of "no surprises" would lessen pressure for hasty action and make it easier for both sides to respond in a calculated and deliberate manner. In time, habits of prior consultation could provide a framework for last-minute efforts to avoid a U.S.-Soviet confrontation or possibly even ad hoc cooperation to head off a local nuclear clash between new nuclear powers.

Nonetheless, both the Soviet Union and the United States in all probability will be reluctant in advance to do more than agree in principle to use their diplomatic good offices and other leverage to defuse crises before they erupt into local nuclear conflict. The divergent interests and perspectives of the Soviet Union and the United States—as well as the complexities of regional disputes—are likely to preclude agreement on prior joint initiatives to reduce the dangers of nuclear war. Further, in some situations, for example, in the Middle East or South Asia, any movement toward such cooperative action would clash with U.S. efforts to reassure its allies threatened by nuclear-armed allies of the Soviet Union.

It would be useful, and perhaps more feasible, to strive for advance agreement on actions that both countries would *not* take for a specified time after detonation of a nuclear weapon in a regional clash. The United States and the Soviet Union might agree, for instance, to defer carrying out past promises of retaliation on behalf of nonnuclear allies; not to ready their air, naval, and ground forces for direct intervention; not to re-supply their ally with military equipment and stocks; not to intervene militarily; and not to indulge in diplomatic posturing and inflammatory behind-the-scenes activity. If prior pledges of retaliation for nuclear use had been made, such a cooling-off period would have to be limited, although to have the desired impact, it would have to be sufficiently long to be more than a formality. Of course, prudence alone would dictate not responding precipitously in a crisis or conflict involving a small nuclear power; but, particularly when valued allies were involved, both the Soviet Union and the United States would be under pressure to take action and would be fearful of what the other would do. Prior agreement that formalized delay would make it easier to follow the dictates of prudence; would help avoid action based on inadequate preliminary reports; and would slow the tempo of events, making it less likely that the conflict would escalate out of control.

Still, there are limits to Soviet-American agreement on any rules of the game for managing their continued competitive involvement in newly nuclearized regions. Soviet activism from the Horn of Africa to Afghanistan suggests that the Soviet leadership is unlikely to subordinate pursuit of its interests— even in newly nuclearized regions. Nor, for that matter, is the United States likely to do so. Besides, the credibility of U.S. pledges of support to nonnuclear countries threatened by Soviet clients would be greatly eroded by too much cooperation with the Soviet Union. Thus, barring a fundamental shock that convinced both countries that more far-reaching restraint was needed, agreement to the principle of consultation, promises of caution, and occasional use of diplomacy to defuse crises where it also would serve other interests might be the most that can be expected. While that would lessen the risk of super-power confrontation, it clearly would not eliminate it.

Defusing Threats to the American Homeland

The United States cannot ignore the possibility that the further spread of nuclear weapons will lead to a threat by a small nuclear power or terrorist group against the American homeland. Because, at least for the 1980s, the most likely threats will rely on unconventional modes of delivery, such as smuggling a bomb into the United States, improved and more timely intelligence is the first line of defense against this type of nuclear threat. For example, the availability of prior warning and detailed information about a terrorist threat to the United States is a precondition to aborting that threat—perhaps by using specially trained forces to seize the weapon overseas or by concentrating U.S. surveillance forces, border patrol units, and customs agents at critical crossing points to preclude smuggling a nuclear weapon into the United States. Monitoring of ships that had docked in ports of potentially hostile countries, or in countries known for harboring terrorists, would reduce the number of vessels that might have to be searched on warning of such a threat. And if a weapon had been smuggled into this country, timely and detailed intelligence about the adversary—in conjunction with improved capabilities for rigorous searching in suspected areas, questioning of suspects, and similar police work—would be the last hope of thwarting attack.

Because of the many difficulties of securing such intelligence, however, defusing a nuclear threat to the American homeland will depend most on successful deterrence. That requires both a reputation for having an intelligence capability sufficient to uncover the identity of an anonymous attacker and the means to retaliate. To deter nuclear attack by a small nuclear power, the United States should threaten to respond by destroying economic, military, political, and industrial targets of high value. Though a limited restructuring of U.S. strategic offensive forces might prove desirable to ensure sufficient range, flexibility, and availability of accurate low-yield warheads, planning for that retaliatory blow would resemble, on a far smaller scale, preparation for deterrence of the Soviet Union. In contrast, if a nuclear attack is carried out by a terrorist group, there might be no clear territorial base against which to retaliate, or the target might be so interspersed with civilian population as to make a conventional nuclear riposte unthinkable. In that case, deterrence might have to depend on the threat to track down and capture—or eliminate—the group's leaders, even if that requires operations on the fringes of legality.

But both deterrence and defense might fail, resulting in the explosion of a nuclear device—most probably on the order of kilotons or tens of kilotons—within an American city. Civil defense preparations could hold down the damage, destruction, and loss of life after such a limited attack. Plans to enable medical, fire, and police personnel to respond; determination of needed medicine, food, and other relief supplies as well as of locations for their storage; procedures for restoring order; required emergency communication equipment; and appropriate training of civil defense and medical personnel all should be under consideration now. But widespread skepticism about

and derision of civil defense against the immeasurably larger
Soviet attack has resulted in neglect of such more feasible and
modest preparations.

  None of these responses to a nuclear threat to the American
homeland entails the procurement of major new weapons systems. But that situation could change following the unexpectedly rapid diffusion of ballistic missile delivery systems to advanced developing countries in the next decade or the
considerably less likely eventuality of a complete breakdown of
the proliferation pattern of the first decades. At that time, serious consideration may have to be given to procurement of a
ballistic missile defense system capable of precluding damage
to the American homeland from small-power nuclear attacks.
But, at least for now, there is little need for such a system.

Lessening Repercussions at Home

Attempting to thwart what intelligence or police sources believe may be preparations for an unconventional nuclear attack
on the American homeland is likely to make it necessary at
times to adopt extraordinary investigatory and police procedures in conflict with traditional American civil liberties. If an
attack is assumed or claimed to be imminent, a series of measures may need to be taken: area-wide and warrantless searches,
the use of informants, illegal or warrantless wiretaps, secret detention, and limited press censorship. Although the actual
threat to accepted procedures and underlying liberal democratic values could prove less serious than sometimes feared,
various steps still should be taken to lessen the chances that any
special measures adopted to guard against unconventional nuclear attack would spill over into other areas of law enforcement.

  For example, administrative regulations governing the taking of such extraordinary measures ought to be drawn very
tightly in order to restrict the situations in which more flexible
practices are to be tolerated. Rigorous enforcement of these
regulations would be an important protection. Judicial review
of those instances in which it is necessary to take steps in tension with traditional civil liberties also would restrain spillovers
and reassert basic values. Further, statements by government

officials, not least the President, that emphasize the extraordinary character of any such infringements and reassert the importance of these liberal democratic values also would serve to check the corrosive impact of exceptional police measures.

There is a danger that if the U.S. were faced with a threat of terrorist or small-power nuclear attack, or in the wake of such an attack, its mood would become xenophobic. There might be outcries for expelling all or selected groups of foreigners, closing U.S. ports to ships from particular countries believed to aid and abet terrorism, banning visitors from those countries, and so on. There might even be violent attacks on the nationals of the country in question. It would fall on elected public officials—and once again especially the President—to resist such a trend by vetoing excessive legislation and using the moral authority of public office to induce restraint. And while it is unlikely that this heightened xenophobia would rekindle now dormant isolationism, policymakers will have to be ready to argue that continued, if more cautious, involvement is still justified in these newly nuclearized regions.

Assessing the Measures

While many of the regional and global threats for which the preceding measures are designed may appear far off in time, events anticipated for the relatively distant future sometimes take place in surprisingly short order. Besides, it is necessary to act now to clarify critical future choices and to identify steps that can be taken to permit more effective responses later. In beginning to think along these lines, however, it is vital to remember that these measures to mitigate the regional and global consequences of proliferation cannot completely neutralize the threats they address and therefore do not provide a fully satisfactory long-term response. Consequently, initiatives to check proliferation's scope, pace, and level, thereby lessening the dangers with which policy must deal, are of critical and continuing importance. For barring a fundamental shock—and quite possibly even then—more far-reaching measures for avoiding nuclear conflict are likely to be unattainable.

AFTER NEXT USE

In the more than three decades that have passed since nuclear weapons were used, both governments and individuals have come to terms with the continuing threat of awesome nuclear destruction—often by subconsciously blocking out the possibility that another use of nuclear weapons could occur. Use of nuclear weapons by a terrorist group or by a small nuclear power in a conflict-prone region would shatter that "nuclear incredulity." Whatever may be thought before the event, the world on the next day would be a different place.

In that new climate, the obstacles to agreement on and implementation of far-reaching measures to deal with the threat of periodic small-power nuclear conflict might be reduced considerably. Nevertheless, to have a reasonable hope of success, a proposed initiative still would have to take into account the particular and often conflicting national interests, ideologies, geographies, cultures, traditions, and domestic politics that underlie the logic of competition among nations. Proposals for change, such as complete nuclear disarmament, that require a utopian total transformation of world politics are not realistic. [8] Only one widely discussed proposal—the establishment of a code of nuclear behavior banning the first use of nuclear weapons—appears even remotely likely to meet that requirement.

Banning First Use

First proposed by Herman Kahn more than a decade ago, and since suggested by others in slightly different forms,[9] a ban on the first use of nuclear weapons is regarded by its proponents both as a nonproliferation measure and as the only available restraint on nuclear anarchy should many countries acquire nuclear weapons. According to Kahn's proposal, the United States would declare that, following a limited period of adjustment to permit the buildup of European conventional forces (the threat of nuclear first use has been the crux of NATO's strategy for deterring Soviet attack), it would not use nuclear weapons first and that it expected other nations to adhere to the same restriction. Henceforth, the only legitimate purpose of nuclear weapons would be to deter the use of other nuclear weapons: no cause, no matter how grave, would justify the first use of those weapons. Explicit, or at least tacit, agreement to this ban on first use would be sought from the Soviet Union as well as from the other nuclear powers.

In order to enforce the ban on first use, Kahn proposed a combination of modern-day versions of banishment, outlawry, and lex talionis. Any country that violated the code would be banished from international society. Trade, aid, investment, and other economic linkages with the country would be severed; its right to participate in international bodies such as the International Monetary Fund or the United Nations would be suspended; its students and tourists no longer welcomed; alliance ties revoked; and its airlines and ships banned from landing or docking in other countries. At the same time, the leaders of the country that had used nuclear weapons first would be declared war criminals, with their removal from office a minimal condition for the country's return to international good grace. As the final and strongest enforcement mechanism, Kahn envisioned a punitive nuclear reprisal, proportional in destruction to the initial violation and conforming to the biblical injunction of an "eye for an eye."

While all countries would be expected to participate in the banishment of the transgressing country, the United States and the Soviet Union would have the main responsibility for carrying out the proportional nuclear reprisal—each on behalf of its allies, friends, and clients—and would be required to stand aside to that retaliation by the other. Either or both would be empowered to respond when nonaligned countries had been attacked with nuclear weapons. Other lesser nuclear powers, such as China, would be permitted to enforce the ban on behalf of their allies. If, despite their commitment, the first user were the United States or the Soviet Union, Kahn's scheme would still propose a proportional, tit-for-tat reprisal—while acknowledging that even a limited response could escalate. Finally, though isolated breakdowns when no nuclear power retaliated on behalf of an attacked country would have to be expected, over many decades, Kahn argued, such a code would minimize recourse to nuclear weapons.

If agreed to by the great powers and enforced as proposed, this code of nuclear behavior would go far to prevent periodic nuclear wars even should proliferation be widespread. A credible threat of proportional nuclear reprisal would deter the first use of nuclear weapons by the nuclear powers. It also would reduce incentives for the acquisition of nuclear weapons in the first place by these and other countries. Despite the value of these objectives, however, there are many obstacles to the acceptance and implementation of such a code.

Potential Problems—and a Fundamental Uncertainty

Many countries may be reluctant to support even the banishment and outlawry of the first user of nuclear weapons. Close foreign policy ties with the transgressor, economic dependence on it, shared ideology, and hostility to the attacked party all could suggest turning a blind eye to the first use of nuclear weapons.

More important, even in the aftermath of a small-power nuclear exchange, the United States may be willing to give priority to enforcing the code above all other American foreign policy and national security objectives. For example, the country that had used nuclear weapons first could be an emerging regional power, for example, India or even Iraq, with which the United States was seeking better relations. Or, the first user might be an important U.S. ally—quite possibly Israel or South Korea—that had used nuclear weapons only as a last resort to avert national defeat and destruction. Standing aside to a Soviet nuclear reprisal in such a case would certainly sacrifice concrete current interests on behalf of an abstract principle of future order. As has so often been the case with other systems of collective security—such as the United Nations and before it the League of Nations—other interests could readily take precedence.[10]

The prospect of Soviet adherence to this type of code also is subject to doubt. Standing aside to an American nuclear reprisal against a transgressor of the ban on first use would have considerable political costs. In an era of Soviet activism abroad, it would convey an appearance of weakness and could make allies and clients question the value of alliance ties to the Soviet Union. Moreover, if China does not participate in the code, the Soviet leadership may fear that Chinese charges of a Soviet-American global condominium will impede Soviet efforts to expand its political influence in the Third World. And nothing in the record of Soviet foreign policy suggests that the Soviet Union will sacrifice pursuit of specific national interests on behalf of a structure of international order such as the code proposes.

Further, even assuming that both superpowers adhere in principle to the first-use ban, there would still be uncertainty about whether the other superpower would actually stand aside to its enforcement. That uncertainty would heighten the risk of carrying out the proposed reprisal and provide another reason not to go through with the punitive response. It also would lessen the deterrent impact of the code, as a new nuclear power might conclude that in a dire situation using nuclear weapons first and taking the chance that there might not be a reprisal was its least bad alternative.

The risk of a nuclear counterreprisal by the country that had violated the no-first-use ban in the first place is still another reason why the United States or the Soviet Union could choose not to carry out the punitive nuclear reprisal. Many new nuclear powers would be capable of mounting at least a crude nuclear threat against one or more Soviet or American cities. In fact, some new nuclear powers might seek this capability precisely to reduce U.S. or Soviet willingness to pledge such retaliation on behalf of allies and friends. While this threat could be reduced by better intelligence preparations and border surveillance as well as active defenses, it cannot be wholly eliminated.

Implementing a punitive nuclear reprisal also presents practical, moral, and political problems. For instance, it would not be easy to determine what constituted a proportional response, an eye for an eye. For example, assuming the first use of nuclear weapons was against a capital city, the economic, political, and/or societal significance of the destroyed capital might be far less than that of the capital of the aggressor. In that case, nuclear reprisal against the first user's capital would not be equivalent but excessive. Similarly, it would be difficult to determine a proportional response to a limited use of a few battlefield nuclear weapons that blocked major invasion corridors. Further, such a ban on first use would entail an unending, global commitment by the United States to other countries' security that the American public is probably unwilling to countenance. But the most difficult dilemma of all is posed by the demand of a tit-for-tat response to the first use of nuclear weapons against cities, for the proportional reprisal required by the code would kill hundreds of thousands of people in cold-blooded retaliation to stand up for an abstract principle of order and stability. But many Americans, as well as peoples in other countries, will question the legitimacy of that reprisal and argue that "two wrongs do not make one right."

Finally, the proposed ban on the first use of nuclear weapons would undermine NATO strategy for ensuring Western European political security and deterring a Warsaw Pact attack. NATO now relies on the threat of nuclear first use to compensate for conventional military weakness and to achieve those objectives.[11] For political and economic reasons, there is little chance that the countries of Western Europe would build up their conventional forces to replace the threat of first use. Rather, either West European political accommodation with the Soviet Union or a West German decision to acquire its own nuclear weapons—which in all probability would otherwise have been avoided—is much more likely to result.*

Thus, a number of political questions and technical problems cast doubt on the feasibility and even the initial acceptability of the proposed ban on nuclear first use. But it is impossible to predict how opinion might change following the next use of

*Kahn recognizes this problem and over the years has dealt with it in different ways. His original proposal called for a European nuclear force; now his argument is that West Germany eventually will acquire nuclear wapons anyway. Other proponents argue that adequate conventional forces can be built up.

nuclear weapons. A small-power nuclear war that escalated to the destruction of cities, thereby causing hundreds of thousands, if not millions, of fatalities would be an awesome and sobering demonstration of the dangers of proliferation as well as of the power and responsibility held by those countries that possess nuclear weapons. It undoubtedly would lead to a widespread humanitarian outcry against the acquisition, possession, modernization, and use of nuclear weapons. If that regional nuclear conflict had threatened to entangle the United States directly, appreciation of the limits of partial measures for dealing with proliferation's consequences and the challenge to traditional statecraft would be greatly magnified. A successful nuclear attack on an American city by a terrorist group or a hostile small power probably would generate an intense and outraged demand for new initiatives to impose order on a dangerous world. Indeed, after nearly four decades of nuclear peace, any use of a nuclear weapon—whether a limited surgical preventive strike by a U.S. ally or a nuclear accident damaging only the new nuclear power itself—could trigger a fundamental revulsion and outcry against traditional approaches to the use and control of military power. Consequently, though a ban on nuclear first use appears now to exceed the realm of the possible, there is no way to anticipate with confidence what would and would not be realistic after next use.

Buttressing the Initial Measures

Even if a ban on first use proves unattainable after the next use of nuclear weapons, there might be a greater chance for the successful undertaking by the United States of other less drastic but still significant initiatives for mitigating proliferation's consequences. Because a "last chance" psychology is likely to prevail, a redoubling of traditional nonproliferation efforts of all sorts probably would meet with widespread domestic and international support. New U.S. attempts, with other countries' collaboration, to reduce the underlying sources of conflict and tension in conflict-prone regions might become more feasible. Still other measures might be pursued to moderate and constrain recourse to force in international politics.

Several steps could be taken, for instance, to strengthen at least the current global presumption—reflected in the belief that nuclear weapons are not simply advanced conventional weapons—against the first use of nuclear weapons.[12] In addition to condemning any country that breaks the nuclear taboo, the United States could sever or reduce economic, political, and other ties with the first user. The possibility of agreement between the United States and the Soviet Union to enforce the nuclear first-use ban on a region-by-region or even rivalry-by-rivalry basis could be explored. Any locally initiated regional no-first-use declarations might be encouraged and adhered to by the United States, while a reinvigorated effort could be made to buttress NATO's conventional forces so that the threat of nuclear first use becomes an unstated last resort rather than a critical linchpin of deterrence in Europe.

The time then also might be ripe for strengthening the presumption in favor of superpower cooperation to defuse regional conflicts that could escalate to a local nuclear clash and to a U.S.-Soviet confrontation. The United States might propose, for instance, agreement in principle with the Soviet Union that both superpowers would subordinate pursuit of their national advantage to head off a nuclear clash between small nuclear powers. Even though in the early stages of a local clash there probably would be different interpretations of which actions violated such an accord and some backsliding by all parties once the memories of nuclear use had dimmed, the very need to justify bending that agreement could help blunt the pursuit of competitive interests. This agreement might not only foster last-ditch efforts to avoid a nuclear eruption, but it also could lead to cooperation in resolving underlying regional disputes before they threatened to get out of hand.

The shock of next use also should make it easier to gain support for regional nuclear arms control initiatives. Especially if next use entails devastating attacks on cities—particularly attacks that are the unintended and tragic result of a nuclear weapons accident or unauthorized access—other countries probably would be more willing to consider seriously such measures as limits on the size and characteristics of their nuclear forces, restrictions on deployment, exchanges of technical information, and limits on nuclear weapons activities. But even if the next use of nuclear weapons is more restricted in scope and consequences, it is probable that some countries' receptivity to nuclear arms control measures would be increased. Offering technical support and encouragement, the United States could take advantage of that heightened receptivity to foster needed agreements.

What Price Proliferation?

Because of the weaknesses of the more modest measures for mitigating proliferation's regional and global consequences, more fundamental, but still nonutopian, changes in our ways of living with nuclear weapons are needed. But even after the next use of nuclear weapons, it may be possible only to buttress those measures. A code of nuclear behavior banning first use is likely to be unattainable. If so, even if it is possible to hold the line at a continuation of the first decades' pattern of slow and limited proliferation, the price of living with the bomb's spread may be the breakdown of nuclear peace in the next decades of the nuclear age.

THE TASKS AHEAD

After India's 1974 nuclear test, halting the spread of nuclear weapons became a critical objective of U.S. foreign policy. But, more recently, in established academic circles and the foreign policy community as well as in some quarters of the Reagan Administration, there has been a growing, if still restrained, undercurrent of concern about the wisdom of an activist non-proliferation policy. The significance for American security of checking the bomb's spread is occasionally questioned. It is sometimes suggested that the next nuclear powers will pose a greater threat to the Soviet than to the American homeland. Besides, the pessimists contend that there is little the United States can do to slow the spread of nuclear weapons and that we should come to terms with the inevitability of a world of nuclear powers.

These counsels must be rejected. The spread of nuclear weapons to conflict-prone regions will make it more difficult to protect U.S. interests in those areas, will threaten the international security and domestic well-being of allies and friends of the United States, and will increase significantly the risk of both terrorist access to nuclear weapons and a U.S.-Soviet confrontation. Moreover, the greater vulnerability of the United States to unconventional nuclear attack and the growing dissemination of missile technology already are evening out proliferation's asymmetrical threat to the Soviet and American homelands.

Most important, the United States, working in cooperation with other countries, *can* affect how many countries acquire

176

nuclear weapons. By fostering presumptions about which countries, under what conditions, and at what point in their civilian nuclear energy programs have legitimate reasons to carry out sensitive reprocessing and enrichment activities, the risks inherent in the global use of nuclear energy would be lessened. A reaffirmation, if not strengthening, of the London Nuclear Suppliers Group Guidelines as well as tightened exports controls over nuclear gray market components and materials would go far to avoid an accelerated erosion of technical constraints. And the earlier warning provided by improved intelligence about nuclear gray and black marketing should permit steps to head off efforts to circumvent even more stringent export controls. Above all, initiatives still can be taken to influence the thinking of potential nuclear powers—not only by adding multilateral muscle to the threat of sanctions but to an even greater degree by an appropriate combination of: preserving or extending American security guarantees, supporting nuclear free zones and other confidence-building measures, selectively supplying advanced conventional arms, and undertaking diplomatic initiatives to alleviate disputes that can fuel decisions to "go nuclear."

But other measures also must be readied both to build proliferation firebreaks in response to outcroppings of additional proliferation and to mitigate the adverse regional and global consequences of the deployment of rudimentary nuclear forces in one or more conflict-prone regions. However, any tendency to focus increasingly more attention on these latter measures at the expense of steps to check the bomb's spread must be resisted, for policies to check proliferation's pace and scope are the keystone of this threefold strategy to contain the nuclear genie in the decades ahead.

Nevertheless, pursuit of this strategy will entail difficult choices and dilemmas, since there will sometimes be tensions or inconsistencies between its components. For example, a tempered response after a nuclear test in an attempt to hold down the level of a country's nuclear weapons activities will undermine efforts to enhance the credibility of sanctions elsewhere. Or, assisting a new nuclear power to develop a safer and less war-prone nuclear force will make still other countries less fearful of building nuclear weapons and deploying their own nuclear forces. Simply focusing greater attention on how to contain future proliferation outcroppings and on how to deal with the consequences of additional proliferation can, to a degree, weaken the disincentives to acquiring nuclear weapons.

Sustaining public support for this more complex strategy will add to the difficulties confronting U.S. policymakers. So-called realists will argue that the United States should stop trying to slow the inevitable and instead get on with the shift to a policy of living with proliferation. But nonproliferation zealots will criticize policymakers for even isolated, limited accommodation with new nuclear powers. And the broader public may find it difficult to appreciate the logic behind a strategy that simultaneously encompasses three divergent and sometimes apparently inconsistent initiatives.

How American policymakers explain and justify this threefold proliferation strategy will be critical both to lessening tensions between its individual parts and winning public support. In particular, it must be stressed that in order to deal with all aspects of the ongoing process of nuclear weapons proliferation, new and diverse policies are demanded. And when it is necessary to adopt an initiative in tension with other measures—say, coming to terms on occasion with a nuclear test—it must be made clear that the ultimate goal that not only legitimizes but also requires that choice is how best to contain proliferation and its consequences.

Successful pursuit by the United States of this threefold strategy for containing the nuclear genie also requires the diplomatic support and cooperation of other countries. The readiness of those countries, running the gamut from the Soviet Union to Saudi Arabia, to use their influence with potential new nuclear powers may be especially important to checking the bomb's spread in some regions—for example, the Middle East and South Asia—or, barring that, at least to holding down the level of proliferation there. Similarly, the support of the major industrial countries is needed for measures ranging from tighter exports controls on gray market components and materials to bolstered sanctions against nuclear weapons-related activities. In addition, high-level discussions with the countries of Western Europe, Japan, and the Soviet Union should be initiated to lay the groundwork for coordinated efforts to build proliferation firebreaks after the outcroppings of additional proliferation. And it is not too soon to open a dialogue with the Soviet Union about possible common actions to cope with the consequences of the nuclearization of conflict-prone regions.

There are bound to be clashes between this complex of policies for containing proliferation and other U.S. foreign, domestic, economic, and security objectives. Provision of new or bolstered security guarantees, for example, to reduce incentives to acquire nuclear weapons may clash with the unwillingness of the American public to accept new responsibilities abroad, with the U.S. interest in improved relations with other countries in the region, and with a desire to reduce the risk of entanglement in local wars. Or, pressures on U.S. allies to tighten their nuclear exports controls could hinder needed efforts to gain the support of these countries for modernizing NATO's theater nuclear forces and for an enhanced military capability to defend Western interests in the Persian Gulf.

Striking a balance between these competing objectives will not be easy. Proliferation policy, of course, is but one part of U.S. foreign and national security policy. Nonetheless, both the adverse consequences of more widespread proliferation for U.S. security and interests and the limitations of policies for mitigating those consequences establish a strong presumption for paying a significant price to influence the scope, pace, and level of proliferation. Depending on the specific situation, some domestic grumbling over new commitments abroad, a risk of military involvement in a local dispute, low-key disputes with NATO allies, worsened relations with some Third World countries, and similar costs all might have to be borne.

But, even with a willingness to pay a significant price to contain the nuclear genie, greater realism about what can be achieved by U.S. policy is necessary. Preventing any more countries from joining the nuclear club is not a reasonable goal. Failure to attain it, moreover, threatens to demoralize supporters of nonproliferation efforts and to engender an unwarranted shift of emphasis to "living with proliferation." A more realistic objective would be to do as well in the next decades of the nuclear age as has been done in the past—to hold the line at no more than, and, one hopes, less than, a continuation of the first decades' pattern of slow and limited proliferation. The United States and other countries also must strive to head off the next use of nuclear weapons while mitigating the other regional and global consequences of the bomb's spread. This, however, may be considerably more difficult.

Thus, while the nuclear genie cannot be put back in the bottle from which it was released, it may still be contained. But a return to a "business-as-usual" approach to proliferation, even punctuated by spasms of intense activity, will accelerate the bomb's spread and bring closer the next use of nuclear weapons.

NOTES

CHAPTER 1

1. Margaret Gowing, *Britain and Atomic Energy, 1939–1945* (London: Macmillan & Co. Ltd., 1964), pp. 16–17.
2. Herbert Feis, *The Atomic Bomb and the End of World War II* (Princeton, N.J.: Princeton University Press, 1970), p. 57.
3. Ibid., pp. 45–50, 193–194; Richard G. Hewlett and Oscar E. Anderson, Jr., *The New World, 1939–1946*, vol. 1: *A History of the United States Atomic Energy Commission* (University Park, Pa.: The Pennsylvania State University Press, 1962), pp. 358–359. However, there were scientists in the Manhattan Project who by early 1945 had begun to question whether to use the bomb. Ibid., pp. 399–400.
4. Feis, *The Atomic Bomb*, pp. 194–196; Hewlett and Anderson, *The New World*, pp. 347–351.
5. Arnold Kramish, *Atomic Energy in the Soviet Union* (Stanford, Calif.: Stanford University Press, 1959), pp. 14–33, 63–64, 105–106.
6. Gregg Herken, *The Winning Weapon: The Atomic Bomb in the Cold War, 1945–1950* (New York: Knopf, 1980), pp. 150–191.
7. Adam B. Ulam, *Expansion and Coexistence: The History of Soviet Foreign Policy, 1917–67* (New York: Praeger, 1968), pp. 415–419.
8. Kramish, *Soviet Union*, p. 94.
9. Andrew Pierre, *Nuclear Politics: The British Experience with an Independent Strategic Force 1939–1970* (London: Oxford University Press, 1972), pp. 9–63, 112–144; R. N. Rosecrance, "British Incentives to Become a Nuclear Power," in *The Dispersion of Nuclear Weapons*, ed. R. N. Rosecrance (New York: Columbia University Press, 1964), pp. 49–51, 53–54, 62–63; Margaret Gowing, *Independence and Deterrence: Britain and Atomic Energy, 1945–1952*, vol. 1: *Policy Making* (New York: St. Martin's Press, 1974), pp. 160–193.
10. Pierre, *Nuclear Politics*, pp. 63–67, 73–79; Rosecrance, "British Incentives," pp. 53–58; Gowing, *Independence and Deterrence*, pp. 184–185, 209–219.

11. Rosecrance, "British Incentives," pp. 56–57, 63–64; Gowing, *Independence and Deterrence*, pp. 213–216.

12. Pierre, *Nuclear Politics*, pp. 73–74; Rosecrance, "British Incentives," p. 65; Gowing, *Independence and Deterrence*, pp. 173–174, 233–234.

13. Gowing, *Independence and Deterrence*, pp. 209, 160–161; Rosecrance, "British Incentives," pp. 58–62; Pierre, *Nuclear Politics*, p. 5.

14. Lawrence Scheinman, *Atomic Energy Policy in France Under the Fourth Republic* (Princeton, N.J.: Princeton University Press, 1965), pp. 64–69, 83–85, 103–125, 181–185; Wolf Mendl, *Deterrence and Persuasion: French Nuclear Armament in the Context of National Policy, 1945–1969* (New York: Praeger, 1970), pp. 123–140.

15. Scheinman, *Fourth Republic*, pp. xxii–xxiii, 119, 167–175, 194, 215–219; Mendl, *Deterrence and Persuasion*, pp. 40–42, 105, 204–205; Wilfrid L. Kohl, *French Nuclear Diplomacy* (Princeton, N.J.: Princeton University Press, 1971), pp. 6–9, 36, 40–41; Pierre, *Nuclear Politics*, pp. 213–214.

16. Scheinman, *Fourth Republic*, pp. 167–173, 215–229; Mendl, *Deterrence and Persuasion*, pp. 21–39, 95–109, 224–225; Kohl, *French Nuclear Diplomacy*, pp. 32–36.

17. Edgar S. Furniss, Jr., *De Gaulle and the French Army* (New York: The Twentieth Century Fund, 1964), pp. 4–5. Furniss's account was confirmed in a conversation with this author by a high-ranking French officer who had been told as much by President de Gaulle.

18. Kohl, *French Nuclear Diplomacy*, pp. 157–158; Général d'Armée Ailleret, "Directed Defense," *Survival* 10, no. 2 (February 1968), reprinted from *Revue de défense nationale* (December 1967), p. 42.

19. Scheinman, *Fourth Republic*, pp. 219–220.

20. Morton H. Halperin, *China and the Bomb* (New York: Praeger, 1965), pp. 71–78.

21. Ibid., pp. 25–26, 44–49; Jonathan D. Pollack, "China as a Nuclear Power," in *Asia's Nuclear Future*, ed. William H. Overholt (Boulder, Colo.: Westview Press, 1977), pp. 37–39.

22. Pollack, "China as a Nuclear Power," pp. 43–44.

23. Ibid., pp. 36–37; Halperin, *China and the Bomb*, pp. 49–53.

24. Leonard Beaton and John Maddox, *The Spread of Nuclear Weapons* (London: Chatto and Windus, 1962), pp. 98–108.

25. Catherine McArdle Kelleher, *Germany and the Politics of Nuclear Weapons* (New York: Columbia University Press, 1975), p. 26.

26. Beaton and Maddox, *The Spread of Nuclear Weapons*, pp. 108–120; Horst Menderhausen, "Will West Germany Go Nuclear?" *ORBIS* 16, no. 2 (Summer 1972), pp. 411–434; Kelleher, *Germany and the Politics of Nuclear Weapons*, pp. 9–32.

27. See Herbert Passin, "Nuclear Arms and Japan," in *Asia's Nuclear Future*, ed. Overholt, pp. 67–132; George Quester, *The Politics of Nuclear Proliferation* (Baltimore: The Johns Hopkins University Press, 1973), pp. 117–121; Yoshiyasu Sato, "Japan's Response to Nuclear Developments: Beyond 'Nu-

clear Allergy,' " in *Nuclear Proliferation and the Near-Nuclear Countries,* ed. Onkar Marwah and Ann Schulz (Cambridge, Mass.: Ballinger, 1975), pp. 227–254.

28. Herbert York, quoted by Deborah Shapley, "Nuclear Weapons History: Japan's Wartime Bomb Projects Revealed," *Science* 199 (January 13, 1978), p. 155.

29. Jan Prawitz, "Sweden—a Non-Nuclear Weapon State," in *Security, Order, and the Bomb,* ed. Johan Jorgen Holst (Oslo: Universitetsforlaget, 1972), pp. 61–66; Beaton and Maddox, *The Spread of Nuclear Weapons,* pp. 151–159.

30. Beaton and Maddox, *The Spread of Nuclear Weapons,* pp. 160–167.

31. This information is based on several confidential personal communications to the author between 1977 and 1981. Also see the off-the-record Yugoslav statement provoked by disputes with the United States over nuclear exports in 1977, which hints at such prior activities, *Nucleonics Week,* April 28, 1977, p. 6.

32. Shapley, "Japanese Bomb," pp. 152–157.

33. David Irving, *The German Atomic Bomb* (New York: Simon and Schuster, 1967), p. 42.

34. Albert Speer, *Inside the Third Reich* (New York: Macmillan, 1970), p. 227.

35. Irving, *The German Atomic Bomb,* pp. 265, 297.

36. John R. Redick, *Military Potential of Latin American Nuclear Energy Programs* (Beverly Hills, Calif.: Sage Publications, 1972), p. 12.

37. This attempt is reported in, among other sources, Mohammed Heikal, *The Road to Ramadan* (New York: Ballantine Books, 1975), pp. 70–71.

38. Robert E. Osgood and Robert W. Tucker, *Force, Order, and Justice* (Baltimore: The Johns Hopkins University Press, 1967), p. 156.

39. For a discussion of the rules of the superpower game, see Osgood and Tucker, *Force, Order, and Justice,* pp. 148–150.

40. For a discussion of the Korean War experience, see Bernard Brodie, *War and Politics* (New York: Macmillan, 1973), pp. 63–65.

41. On this theme, see ibid., pp. 392–416; Michael Mandelbaum, *The Nuclear Question* (New York: Cambridge University Press, 1979), pp. 124–128, 134–144.

42. A classic discussion of the requirements of deterrence remains. Albert Wohlstetter, "The Delicate Balance of Terror," *Foreign Affairs* 37, no. 2 (January 1959). For another insightful discussion of the "technical" requirements of stable deterrence, see Thomas C. Schelling, *Arms and Influence* (New Haven: Yale University Press, 1966), pp. 221–248; Fred Charles Iklé, "Can Deterrence Last Out the Century?" *Foreign Affairs* 51, no. 2 (January 1973).

43. For a cogent critique of launch-on-warning, see Paul Wolfowitz, "The Proposal to Launch on Warning," in U.S., Senate, Committee on Armed Services, *Authorization for Military Procurement, Research and Development, Fiscal Year 1971, and Reserve Strength, Hearings,* Pt. 3, 91st Cong., 2d Sess., 1970, pp. 2278–2282.

44. Some instances of accidents involving nuclear weapons in which the safety measures prevented nuclear explosion are detailed in Clyde Burleson, *The Day the Bomb Fell on America: True Stories of the Nuclear Age* (Englewood Cliffs, N.J.: Prentice-Hall, 1978), pp. 1–17; Joel Larus, *Nuclear Weapons Safety and the Common Defense* (Columbus: Ohio State University Press, 1967, pp. 86–99.

45. Donald G. Brennan, one person so authorized to raise these issues, related this to the author. A primary channel of communication was a speech at Ann Arbor, Michigan, in 1962 by then Assistant Secretary of Defense John McNaughton, setting out many command, control, and safety concepts. See John T. McNaughton, "Arms Restraint in Military Decisions," December 19, 1962, mimeographed.

46. Herman Kahn, Lecture at Hudson Institute, Croton-on-Hudson, New York, November 6, 1978.

47. See Iklé, "Can Deterrence Last Out the Century?" pp. 272–276.

CHAPTER 2

1. For a discussion of the role of the United States in the spread of this basic theoretical knowledge, see Clarence Long, "Nuclear Proliferation: Can Congress Act in Time?" *International Security* 1, no. 4 (Spring 1977), pp. 54–63.

2. See Albert Wohlstetter et al., *Moving Toward Life in a Nuclear Armed Crowd?* Pan Heuristics, report prepared for the U.S. Arms Control and Disarmament Agency, April 22, 1976, pp. 14, 249–261.

3. "Iraq Said to Get A-Bomb Ability with Italy's Aid," *New York Times*, March 18, 1980.

4. *Nucleonics Week*, October 27, 1977, pp. 1–2.

5. Ted Greenwood, George W. Rathjens, and Jack Ruina, *Nuclear Power and Weapons Proliferation*, Adelphi Paper no. 130 (London: The International Institute for Strategic Studies, 1977), p. 18; also see John R. Lamarsh, "Dedicated Facilities for the Production of Nuclear Weapons in Small and/ or Developing Nations," in *Nuclear Proliferation and Safeguards*, app. vol. 2, pt. 2 (Washington, D.C.: Office of Technology Assessment, 1977) app. 4–A, p. 27; Wohlstetter et al., *Moving Toward Life in a Nuclear Armed Crowd?* pp. 200–208; William Van Cleave, "Nuclear Technology and Weapons," in *Nuclear Proliferation Phase II*, ed. Robert M. Lawrence and Joel Larus (Lawrence: University Press of Kansas, 1974), pp. 46–50; Report by the Comptroller General of the United States, "Quick and Secret Construction of Plutonium Reprocessing Plants: A Way to Nuclear Weapons Proliferation?" (Washington, D.C.: General Accounting Office, October 6, 1978), pp. 5–10.

6. See Lamarsh, "Dedicated Facilities"; Greenwood et al., *Nuclear Power and Weapons Proliferation*, pp. 18–19, 29; Gene I. Rochlin, "The Development

and Deployment of Nuclear Weapons Systems in a Proliferating World,"
in *International Political Effects of the Spread of Nuclear Weapons,* ed. John
Kery King (Washington, D.C.: U.S. Government Printing Office, 1979).
7. See Greenwood et al., *Nuclear Power and Weapons Proliferation,* pp. 21–28.
8. For more detail, see ibid., pp. 3–6; Van Cleave, "Nuclear Technology and
Weapons," pp. 51–55; John McPhee, *The Curve of Binding Energy* (New
York: Farrar, Straus and Giroux, 1974). McPhee's *The Curve of Binding Energy* is a highly readable discussion of these issues in the form of a profile
of Ted Taylor who, during the 1940s and 1950s, was a designer of nuclear
weapons at the then Atomic Energy Commission's Los Alamos facility.
9. Van Cleave, "Nuclear Technology and Weapons," pp. 53–54; Frank C.
Barnaby, "How States Can 'Go Nuclear,' " *The Annals* (March 1977), p. 40.
Their analyses were confirmed by discussions with the late Donald G.
Brennan, Director of National Security Studies, Hudson Institute, Croton-
on-Hudson, New York.
10. On growing indigenous capabilities for aircraft production, see "1980 Aerospace Forecast and Inventory," *Aviation Week and Space Technology,* March
3, 1980, pp. 120–123. For an up-to-date listing of world aircraft invento-
ries, see *The Military Balance, 1979–1980* (London: The International Institute for Strategic Studies, 1980).
11. On Israeli missile development and the presence of surface-to-surface missiles in the Middle East, see Robert E. Harkavy, *Spectre of a Middle Eastern
Holocaust. The Strategic and Diplomatic Implications of the Israeli Nuclear Weapons Program,* Monograph Series in World Affairs, University of Denver,
Graduate School of International Studies, pp. 33–36; *The Military Balance,*
passim. On South Korea and Taiwan, see, respectively, *Aviation Week and
Space Technology,* October 2, 1978, p. 11, and "Propping Up a Fading
Friendship," *Far Eastern Economic Review,* October 27, 1978, p. 18.
12. Rochlin, "The Development and Deployment of Nuclear Weapons Systems
in a Proliferating World," pp. 15–16; on the more general point of available delivery systems, see also pp. 12–17.
13. See John E. Endicott, *Japan's Nuclear Option* (New York: Praeger, 1975);
"1980 Aerospace Forecast and Inventory," *Aviation Week and Space Technology,* March 3, 1980, pp. 107–109, 111, 149.
14. Onkar Marwah, "India's Nuclear and Space Programs: Intent and Policy,"
International Security 2, no. 2 (Fall 1977); "India Launches New Satellite,"
Aviation Week and Space Technology, July 28, 1980, p. 17.
15. R. B. Murrow, "Nuclear Proliferation: Readymade Aerial Delivery Possibilities for Lesser Nations," The RAND Corporation, RM–4806–PR/ISA,
March 1966 (declassified March 1978).
16. Mohammed Heikal, *The Road to Ramadan* (New York: Ballantine Books,
1975), pp. 70–71. See also reports in "Pakistan Denies It Plans A-Bomb;
Denounces Washington Aid Cutoff," *New York Times,* April 9, 1979; "Panel
Told Pakistan Gained A-Weapons Ability by 'End-Runs,' " *Washington Post,*
May 2, 1979; "Is Qaddafi Financing Pakistan's Nuclear Bomb?" *Christian
Science Monitor,* December 19, 1979.

17. See, for example, "Panel Told Pakistan Gained A-Weapons Ability by 'End-Runs,' " *Washington Post,* May 2, 1979; Zalmay Khalilzad, "Pakistan and the Bomb," *Survival* 21, no. 6 (November–December 1979), p. 247; BBC-TV Panorama Investigative Team, "The Birth of the Islamic Bomb," photostat distributed by the New York Times Syndicate Sales Corp., 1980.

18. *Nucleonics Week,* June 26, 1980; "Israeli Planes Bomb Major Iraqi Nuclear Facility," *Washington Post,* June 9, 1981.

19. "L'Agence Internationale de l'Energie Atomique approuve les accords nucléaires Germano-Bresiliens et Franco-Pakistanais," *Le Monde* (Paris), February 28, 1976; Norman Gall, "Atoms for Brazil, Dangers for All," *Foreign Policy,* no. 23 (Summer 1976), pp. 155–158.

20. "Pakistan Says France Killing Controversial Nuclear Deal," *Washington Post,* August 24, 1978.

21. "The Nuclear Rivalries in Latin America," *Financial Times* (London), April 6, 1979.

22. See "Nuclear Suppliers Group, Guidelines for Nuclear Transfers," transmitted to the International Atomic Energy Agency, January 1978, reprinted in Atlantic Council, *Nuclear Power and Nuclear Weapons Proliferation,* Report of the Atlantic Council's Nuclear Fuels Policy Group (Washington, D.C.: The Atlantic Council, 1978), vol. 11, app. D, pp. 63–75.

23. President Jimmy Carter, "Remarks of the President on Nuclear Power Policy and Question and Answer Session," April 7, 1977 (Washington, D.C.: Office of the White House Press Secretary, 1977).

24. See *INFCE Summary Volume* (Vienna, Austria: International Atomic Energy Agency, 1980).

25. See Nuclear Non-Proliferation Act of 1978, Public Law 95–242, March 10, 1978. Its main provisions, including those not touched on here, are summarized in *Nuclear Power and Nuclear Weapons Proliferation,* vol. 11, app. B, pp. 30–33.

26. On all these various aspects of the Argentine–Swiss–German deal, see *Nucleonics Week,* November 29, 1979, p. 5; April 3, 1980, pp. 3–4; April 17, 1980, p. 4; "Canada and Germany Fall Out over Atucha," *Nuclear Engineering International,* January 1980, p. 10; "U.S. Tries to Prevent Swiss Sale to Argentina for Atom Program," *New York Times,* March 11, 1980; "Argentina Is Nearing Deal for Reactor that Makes Plutonium," *New York Times,* March 27, 1980; "U.S.-Opposed European Nuclear Sale to Argentina Awaited," *Washington Post,* February 10, 1980; "U.S. Fails to Halt German Reactor for Argentina," *Washington Post,* April 4, 1980; "Swiss to Build Argentina N-Plant," *Financial Times* (London), March 15, 1980; "Argentina Gets into Heavy Water," *Latin America Weekly Report,* March 21, 1980; "KWU Wins Argentine Nuclear Power Contract," *Latin America Economic Report,* October 5, 1979; "Swiss Will Build Heavy Water Plant," *World Business Weekly,* March 31, 1980.

27. Testimony of Assistant Secretary of State Thomas C. Pickering before the Committee on Government Affairs, Subcommittee on Energy, Nuclear Proliferation, and Federal Services, U.S., Senate, May 1, 1979 (photostat);

Khalilzad, "Pakistan and the Bomb," p. 247; *Nucleonics Week,* April 12, 1979, pp. 4–5; June 28, 1979, p. 11; July 5, 1979, pp. 4–5; March 6, 1980, p. 1; "U.S. Aid to Pakistan Cut after Evidence of Atom Arms Plan," *New York Times,* April 7, 1979; "Pakistan Denies It Plans A-Bomb; Denounces Washington Aid Cutoff," *New York Times,* April 9, 1979; "Gate Shuts on Pakistan's Nuclear Path," *Financial Times* (London), April 11, 1979; "Swiss, U.S. Prepared to Resume Nuclear Cooperation," *Washington Post,* December 31, 1980; "Pakistan Said to Receive Nuclear Arms Parts Illegally via Canada," *Washington Post,* December 7, 1980.

28. See BBC-TV Panorama Investigative Team, "The Birth of the Islamic Bomb," pp. 16–33.
29. See references to such reports in, for example, "Says U.S. Policy Is Shaped by 'Zionist Circles' Who Fear Threat to Israelis," *New York Times,* April 9, 1979; *Nucleonics Week,* April 12, 1979, p. 5; "Nuclear Arms Race: The Field Grows Larger," *The Guardian* (London), January 13, 1980.

See also reports of speculation that, even if a South Asian nuclear weapons–free zone were to be obtained, Pakistan might explode its device in Saudi Arabia, a country neither a member of that zone nor a signatory of the NPT, in *Nucleonics Week,* April 19, 1979, p. 8.
30. See, for example, Harkavy, *Spectre of a Holocaust,* p. 78.
31. See "Israel Denies a Report It Tested Atom Bomb in the South Atlantic," *New York Times,* February 23, 1980; "Israel Penalizes Reporter for Violating Censorship," *Washington Post,* February 25, 1980; "New U.S. Concern: Repercussions over Nuclear-Type Flash," *Christian Science Monitor,* February 25, 1980.
32. "3 Nations to Begin Cruise Missile Project," *Washington Post,* December 8, 1980. Also see Robert Harkavy's discussion of the prospect for nuclear weapons cooperation among a group of "pariah" states, including Israel, South Africa, Taiwan, and South Korea. Robert Harkavy, "The Pariah State Syndrome," *ORBIS* 21, no. 3 (Fall 1977).
33. Ernest W. Lefever, *Nuclear Arms in the Third World: U.S. Policy Dilemma* (Washington, D.C.: The Brookings Institution, 1979), p. 70.
34. "Pakistan Stole Nuclear Secret in the Netherlands," *NRC Handelsblad* (New Rotterdam, Holland), June 16, 1979.
35. *Nucleonics Week,* January 24, 1980, pp. 9–10.
36. *Nucleonics Week,* March 13, 1980, p. 11.
37. For a more detailed discussion, see Mason Willrich and Theodore B. Taylor, *Nuclear Theft: Risks and Safeguards* (Cambridge, Mass.: Ballinger, 1974).
38. "C.I.A. Said in 1974 Israel Had A-Bombs," *New York Times,* January 27, 1978; *Nucleonics Week,* October 13, 1977, p. 7; Lefever, *Nuclear Arms in the Third World,* p. 65.
39. "U.S. Acknowledges Possibility of a Uranium Theft," *New York Times,* April 28, 1979. See also *Nucleonics Week,* June 14, 1979, p. 7; "Testimony Doubted on Missing Uranium," *New York Times,* April 28, 1979; "U.S. Unravels Apollo's Losses of Nuclear Material," *Washington Star,* June 26, 1977.
40. *Nuclear Engineering International,* June 1977, p. 6; Elaine Davenport, Paul

Eddy, and Peter Gillman, *The Plumbat Affair* (Philadelphia: J. B. Lippincott, 1978). "Plumbat" is reported to have been the Israeli code name for the hijacking operation.

41. Department of Atomic Energy, *Annual Report, 1979–80*, p. 52.
42. On this deal, see, for example, "Nuclear Take-Off," *Latinamerican Week,* March 25, 1977, p. 6; "Nuclear Diplomacy," *Latinamerican Week,* January 13, 1978, p. 7; "Nuclear Knowhow for Sale," *Latin America Economic Report* 6, no. 34, September 1, 1978; "Peru Seeks to Boost Trade with Argentina," *Latin America Economic Report* 7, no. 23, June 15, 1979, p. 178.
43. See "Nuclear Diplomacy," *Latinamerican Week,* January 13, 1978; "Nuclear Knowhow for Sale," *Latin America Economic Report* 6, no. 34, September 1, 1978; "Argentine: Accord de coopération nucléaire avec la Bolivie," *Défense et Diplomatie,* April 27, 1978; "Amérique Latine: L'OEA et l'atome," *Défense et Diplomatie,* September 10, 1979; "Argentine/Bresil: Coopération ou compétition atomique?" *Défense et diplomatie,* March 3, 1980.
44. *Nucleonics Week,* October 11, 1979, p. 3.
45. "Text of Brazilian-Iraqi Nuclear Agreement," *O Estado de São Paulo,* January 25, 1980, reprinted in Foreign Broadcast Information Service, *Worldwide Report: Nuclear Development and Proliferation,* no. 33, JPRS 75245, March 4, 1980, pp. 5–6; "Brazilian Minister Confirms Nuclear Agreement with Iraq," *O Estado de São Paulo,* November 25, 1979, reprinted in Foreign Broadcast Information Service, *Worldwide Report: Nuclear Development and Proliferation,* no. 24, JPRS 74890, January 9, 1980, p. 4; *Nucleonics Week,* November 15, 1979, p. 1; January 17, 1980, p. 10.
46. *World Business Weekly,* August 6–12, 1979; "Brazil, Chile Nuclear Negotiations," *O Estado de São Paulo,* March 6, 1980, reprinted in Foreign Broadcast Information Service, *Worldwide Report: Nuclear Development and Proliferation,* no. 37, JPRS 75431, April 3, 1980, p. 26.

   In the author's discussion with Brazilian government officials in early 1980, they discussed Brazil's intentions to extend its nuclear dealings into these regions in the years ahead.
47. *Nucleonics Week,* April 17, 1980, pp. 4–5; "Madero: Brazil and Argentina Will Respect Nuclear Standards," *O Estado de São Paulo,* February 15, 1980, reprinted in Foreign Broadcast Information Service, *Worldwide Report: Nuclear Development and Proliferation,* no. 35, JPRS 753222, March 17, 1980, pp. 15–16; "Argentina to Provide Pipe," *Jornal do Brasil,* February 15, 1980, reprinted in Foreign Broadcast Information Service, *Worldwide Report: Nuclear Development and Proliferation,* no. 35, JPRS 753222, March 17, 1980, pp. 16–17.
48. "Nuclear Training Offer to Developing Countries," *Korea Times,* November 22, 1979.
49. India and Libya, for example, had a falling out in 1979 because of India's unwillingness to supply Libya with nuclear assistance and technology that had military applications. *Nucleonics Week,* August 30, 1979, p. 10; "Libya Pressures India to Supply Nuclear Technology," *Financial Times* (London),

September 1, 1979. These reports were confirmed by the author in discussions in New Delhi in 1981.

50. This draws on discussions with Brazilian officials soon after the deal was announced.

51. "Sale of Sensitive Technology to Iraq Denied; Comments," *O Estado de São Paulo*, January 12, 1980, reprinted in Foreign Broadcast Information Service, *Worldwide Report: Nuclear Development and Proliferation*, no. 29, JPRS 75122, February 12, 1980, pp. 25–26.

52. "Statements by Castro Madero," *O Estado de São Paulo*, January 30, 1980, reprinted in Foreign Broadcast Information Service, *Worldwide Report: Nuclear Development and Proliferation*, no. 32, JPRS 75201, February 26, 1980, p. 9. See also *Nucleonics Week*, December 20, 1979, p. 8; April 17, 1980, pp. 4–5.

CHAPTER 3

1. BBC-TV Panorama Investigative Team, "The Birth of the Islamic Bomb," (photostat distributed by the New York Times Syndicate Sales Corp., 1980), pp. 3–7.

2. "Pakistan Close to Full Nuclear Capability," *Financial Times* (London), October 5, 1978. See also "U.S. Officials View Pakistan as the Leading Threat to Join the Nuclear Club," *Washington Post*, December 8, 1978.

3. BBC-TV Panorama Investigative Team, "The Birth of the Islamic Bomb," pp. 16–33.

4. *Nucleonics Week*, October 12, 1978, p. 13.

5. *Nucleonics Week*, July 5, 1979, pp. 4–5; "Pakistan Stole Nuclear Secret in the Netherlands," *NRC Handelsblad* (New Rotterdam, Holland), June 16, 1979; "UK Deals May Aid Pakistan Atom Bid," *Financial Times* (London), August 22, 1979.

6. "Fears Rising in Washington That an India-Pakistan Nuclear Arms Race Is Inevitable," *New York Times*, August 24, 1979.

7. Robert Shaplen, "Profiles (David Newsom—Part III)," *The New Yorker*, June 16, 1980, p. 65; Shirin Tahir-Kheli, "Pakistan's Nuclear Option and U.S. Policy," *ORBIS* 22, no. 2 (Summer 1978), pp. 361–362; Zalmay Khalilzad, "Pakistan and the Bomb," *Survival* 21, no. 6 (November–December 1979), pp. 245–248.

8. See also Richard K. Betts, "Incentives for Nuclear Weapons: India, Pakistan, Iran," *Asian Survey* 19, no. 11 (November 1979), pp. 1060–1061; Tahir-Kheli, "Pakistan's Nuclear Option and U.S. Policy," p. 361.

9. Tahir-Kheli, "Pakistan's Nuclear Option and U.S. Policy," p. 357; "Pakistan Drafts a Plan for Nuclear Development," *Christian Science Monitor*, December 12, 1977.

10. Quoted by Tahir-Kheli, "Pakistan's Nuclear Option and U.S. Policy," p. 367.

11. See "Pakistan Drafts a Plan for Nuclear Development," *Christian Science Monitor*, December 12, 1977; Betts, "Incentives for Nuclear Weapons," p. 1071.

12. Statement by Senator Alan Cranston, reported in "Cranston Says India and Pakistan Are Preparing for Nuclear Testing," *New York Times*, April 28, 1981.

13. Paul F. Power, "The Indo-American Nuclear Controversy," *Asian Survey* 19, no. 6 (June 1979), pp. 582–583.

14. Press release from Delhi Domestic Service, March 13, 1980, reprinted in Foreign Broadcast Information Service, *Worldwide Report: Nuclear Development and Proliferation*, JPRS 75431, no. 37, April 3, 1980, p. 21, also see *Nucleonics Week*, April 2, 1981.

15. Statement by Senator Alan Cranston, reported in "Cranston Says India and Pakistan are Preparing for Nuclear Testing," *New York Times*, April 28, 1981.

16. Stephen P. Cohen, "Perception, Influence, and Weapons Proliferation in South Asia" (Report prepared for the Department of State, Bureau of Intelligence and Research/External Research, Contract no. 1722–920184, August 20, 1979), p. 41. More generally on Indian proliferation incentives, see Ashok Kapur, *India's Nuclear Option: Atomic Diplomacy and Decision Making* (New York: Praeger, 1976); Shelton L. Williams, *The U.S., India and the Bomb* (Baltimore, Md.: The Johns Hopkins University Press, 1969); Onkar Marwah, "India's Nuclear and Space Programs: Intent and Policy," *International Security*, vol. 2, no. 2 (Fall 1977); Betts, "Incentives for Nuclear Weapons," pp. 1054–1058.

17. On that claim, see Stephen P. Cohen and Richard L. Park, *India: Emergent Power?* (New York: Crane, Russak, 1978), pp. 4–5.

18. On these linkages with Chinese activities, see Kapur, *India's Nuclear Option*, pp. 232–244; Betts, "Incentives for Nuclear Weapons," pp. 1055–1058; Cohen, "Perception, Influence, and Weapons Proliferation in South Asia," pp. 44–45.

19. For still germane expressions of these undercurrents, see Indira Gandhi, "India and the World," *Foreign Affairs* 51, no. 1 (October 1970); Baldev Raj Naijar, "Treat India Seriously," *Foreign Policy*, no. 18 (Spring 1975), pp. 134–154.

20. Marwah, "India's Nuclear and Space Programs," pp. 106–113.

21. Cohen, "Perception, Influence, and Weapons Proliferation in South Asia," p. 41.

22. "C.I.A. Said in 1974 Israel Had A-Bombs," *New York Times*, January 27, 1978.

23. See Robert E. Harkavy, *Spectre of a Middle Eastern Holocaust: The Strategic and Diplomatic Implications of the Israeli Nuclear Weapons Program*, Monograph Series in World Affairs (University of Denver, Graduate School of International Studies), pp. 5–19.

24. "C.I.A. Said in 1974 Israel had A-Bombs," *New York Times*, January 27, 1978.

25. "Dayan Says Israeli's Have the Capacity to Produce A-Bombs," *New York Times*, June 25, 1981.

26. Yair Evron, "Israel and the Atom: The Uses and Misuses of Ambiguity, 1957–1967," *ORBIS* 17, no. 4 (Winter 1974), pp. 1330–1332; Avigdor Haselkorn, "Israel: From an Option to a Bomb in the Basement?" in *Nuclear Proliferation Phase II*, ed. Robert M. Lawrence and Joel Larus (Lawrence: University Press of Kansas, 1974), pp. 166–173.

27. On the various rationales for Israel's nuclear weapons program, see Harkavy, *Spectre of a Holocaust*, pp. 57–80.

28. "U.S. Taking Long Gamble on Sadat," *Washington Post*, March 30, 1980; "U.S. Military Sales to Saudis 5 Times Total for Israelis," *Washington Post*, October 11, 1979; "Pentagon Plans to Supply Saudis with Bombs and Missiles for U.S. F5s." *Washington Post*, December 12, 1979.

29. Paul Rivlin, "The Burden of Israel's Defense," *Survival* 20, no. 4 (July–August 1978), pp. 146–154.

30. "Israeli Chief of Staff Says Iraq Is Developing Nuclear Weapons," *New York Times*, July 30, 1977; "The Problems Go on Mounting for an Unsure, Worried Israel," *World Business Weekly*, March 17, 1980, pp. 22–23.

    On U.S. views, see "Concern Increasing on A-Bomb's Spread," *New York Times*, April 7, 1980; "Iraq A-Bomb Seen by 1985," *International Herald Tribune* (London), June 27, 1980.

31. "Iraq Asserts Arabs Must Acquire Atom Arms as a Balance to Israel," *New York Times*, June 24, 1981.

32. "France Plans to Sell Iraq Weapons-Grade Uranium," *Washington Post*, February 28, 1980; "France Cites Nuclear Terms to Iraq," *Washington Post*, July 30, 1980.

33. "Iraq Said to Get A-Bomb Ability with Italy's Aid," *New York Times*, March 18, 1980; "Italian Aides Deny Iraq Gained A-Bomb Ability," *New York Times*, March 19, 1980.

34. ABC News Closeup, "Near Armageddon: The Spread of Nuclear Weapons in the Middle East," April 27, 1981.

35. *Nucleonics Week*, November 15, 1979, p. 1; January 17, 1980, p. 10; "Sale of Sensitive Technology to Iraq Denied; Comments," *O Estado de São Paulo*, January 12, 1980, reprinted in Foreign Broadcast Information Service, *Worldwide Report: Nuclear Development and Proliferation* no. 29, JPRS 75122, February 12, 1980, pp. 25–26.

36. "Portugal Sells Uranium to Iraq," *Financial Times* (London), March 28, 1980; "And the Nuclear Race Goes On," *Newsweek*, June 22, 1981.

37. *Nucleonics Week*, August 30, 1979, p. 10; "Libya Presses India on Nuclear Technology," *Nuclear Engineering International*, November 1979, pp. 7–8; "Libya Pressures India to Supply Nuclear Technology," *Financial Times* (London), September 1, 1979.

38. BBC-TV Panorama Investigative Team, "The Birth of the Islamic Bomb," pp. 9–12; Khalilzad, "Pakistan and the Bomb," p. 248. Not all U.S. officials agree, however. See "More Nations Elbow Way into World's Nuclear Club," *Christian Science Monitor*, June 16, 1980.

39. "Qadhafi Move Sparks New Nuclear Fears," *International Herald Tribune* (London), January 21, 1981.
40. On Iraq's new foreign policy, see Claudia Wright, "Iraq New Power in the Middle East," *Foreign Affairs* 58, no. 2 (Winter 1979–80), pp. 257–277.
41. "Pride and Punishment," *New York Times,* June 11, 1981.
42. Ibid.
43. "U.S.-Soviet Exchange about South Africa Said to Improve Ties," *New York Times,* August 29, 1977; "Soviet Flap about South African Bomb," *Christian Science Monitor,* August 11, 1977.
44. "U.S.-Soviet Exchange about South Africa Said to Improve Ties," *New York Times,* August 29, 1977; "The Enriched Uranium Route," *Financial Times* (London), August 24, 1977; "United States Asks South Africa about A-Bomb Plan," *New York Times,* August 21, 1977; "South African Stirs New A-Arms Flurry," *New York Times,* August 31, 1977.
45. "President Carter's New Conference of August 23," *Department of State Bulletin,* September 19, 1977, p. 376.
46. "U.S. Disagrees with Vorster on A-Weapons," *Washington Post,* October 25, 1977.
47. Eliot Marshall, "Flash Not Missed by Vela Still Veiled in Mist," *Science* 206, November 30, 1979, p. 1051.
48. Ibid.; "U.S. Officials Uncertain about that Event near South Africa," *Washington Post,* October 27, 1979.
49. "U.S. Monitors Signs of Atomic Explosion near South Africa," *New York Times,* October 26, 1979; Marshall, "Flash Not Missed by Vela Still Veiled in Mist," p. 1051; Eliot Marshall, "Scientists Fail to Solve Vela Mystery," *Science* 207, February 1, 1980, pp. 504–505; "Data Suggesting Bomb Test in South Atlantic Region Revised," *New York Times,* November 27, 1979.
50. Marshall, "Scientists Fail to Solve Vela Mystery," pp. 504–505.
51. Lee Torrey, "Is South Africa a Nuclear Power?" *New Scientist,* July 24, 1980, p. 268; Eliot Marshall, "Navy Lab Concludes the Vela Saw a Bomb," *Science* 209, August 29, 1980, pp. 996–997.
52. On these diplomatic uses, see also Richard K. Betts, "A Diplomatic Bomb for South Africa?," *International Security* 4, no. 2 (Fall 1979), pp. 101–105; Robert S. Jaster, *South Africa's Narrowing Security Options,* Adelphi Papers No. 159 (London: International Institute for Strategic Studies, 1980), pp. 45–46.
53. Kenneth L. Adelman and Albion W. Knight, "Can South Africa Go Nuclear?" *ORBIS* 23, no. 3 (Fall 1979), p. 635; J. E. Spence, "South African Foreign Policy: Changing Perspectives," *The World Today* 34, no. 11 (November 1978), p. 425; Jaster, *South Africa's Narrowing Security Options.* p. 45.
54. Jaster, *South Africa's Narrowing Security Options,* p. 28; William Gutteridge, "South Africa's Defence Posture," *The World Today* 36, no. 1 (January 1980), p. 28.
55. Kenneth L. Adelman, "The Strategy of Defiance: South Africa," *Comparative Strategy* 1, nos. 1 and 2, pp. 46, 50; Spence, "South African Foreign Policy," p. 421.
56. Jaster, *South Africa's Narrowing Security Options,* pp. 27–30; Adelman, "The

Strategy of Defiance," pp. 643–644; Gutteridge, "South Africa's Defence Posture," pp. 28–30.

57. Betts, "A Diplomatic Bomb for South Africa?" pp. 104–105; Jaster, *South Africa's Narrowing Security Options,* p. 45.

58. "Interview with Professor Iya Abubakar, Minister of Defense of Nigeria," *West Africa,* May 19, 1980. See also Eddie Iroh, "Nigeria: The Nuclear Debate," *Africa,* no. 114, February 1981, pp. 38–39.

59. *China Quarterly* 64 (December 1975), p. 808.

60. See Ernest W. Lefever, *Nuclear Arms in the Third World: U.S. Policy Dilemma* (Washington, D.C.: The Brookings Institution, 1979), p. 92; George H. Quester, "Taiwan and Nuclear Proliferation," *ORBIS* 17, no. 1 (Spring 1974), pp. 148–149; Ralph N. Clough, *Island China,* a Twentieth Century Fund Study (Cambridge, Mass.: Harvard University Press, 1978), pp. 116–120.

61. Lefever, *Nuclear Arms in the Third World,* p. 89; "Taipei Treads Lightly," *Far Eastern Economic Review,* January 6, 1978; "Taiwan Forces Reportedly Buy Israeli Missiles," *New York Times,* April 6, 1977; "Reprocessing to Halt, Administration Says," *Nuclear News,* November 1976; "East, West Anxiously Watch Taiwan's Nuclear Progress," *International Herald Tribune* (London), March 2, 1977.

62. "Investigation of Korean-American Relations," Report of the Subcommittee on International Organization, Committee on International Relations, U.S., Congress, House of Representatives, October 31, 1978, pp. 79–80; "U.S. Kept S. Korea from Making A-Arms," *International Herald Tribune* (London), November 6, 1978; "South Korea Drops Plan to Buy a Nuclear Plant from France," *New York Times,* January 30, 1976; "Seoul Officials Say Strong U.S. Pressure Forced Cancellation of Plans to Purchase a French Nuclear Plant," February 1, 1976.

63. "Statement of President Jimmy Carter," July 20, 1979, *Department of State Bulletin,* September 1979, p. 37.

64. "China Steps up Protest of U.S. Arms for Taiwan," *Washington Post,* June 22, 1980.

65. "Half a Loaf Is Better than None," *Far Eastern Economic Review,* January 25, 1980, p. 22; "Weapons Sales to Taiwan Approved," *Aviation Week and Space Technology,* January 14, 1980, p. 12.

66. The following draws on my discussions with Taipei and South Korean officials and academics in Taiwan, in South Korea, and in the United States. Also see Young-sun Ha, "Nuclearization of Small States and World Order: The Case of Korea," *Asian Survey* 18, no. 11 (November 1978), pp. 1141–1144; William H. Overholt, "Nuclear Proliferation in Eastern Asia," in *Asia's Nuclear Future,* ed. William H. Overholt (Boulder, Colo.: Westview Press, 1977), pp. 139–149; Quester, "Taiwan and Nuclear Proliferation," pp. 144–149.

67. U.S., Congress, Senate, Committee on Governmental Affairs, and House, Committee on Foreign Affairs, *Nuclear Proliferation Factbook,* 96th Cong., 2d sess., 1980, p. 325; see also Ha, "Nuclearization of Small States and World Order," pp. 1135–1141.

68. Overholt, "Nuclear Proliferation in Eastern Asia," pp. 153–154, 159.
69. Juan E. Guglialmelli, "And If Brazil Builds an Atomic Bomb?: The Brazilian–German Nuclear Agreement," *Estrategia* (May–June and July–August 1975), reprinted in *Survival* 18, no. 4 (July–August 1976), pp. 164–165.
70. "Nations Thinking Nuclear Worry about Neighbors," *Washington Post*, December 8, 1978; "While Brazil Grapples with German Nuclear Contract . . . ," *Latin American Economic Report*, November 17, 1978.
71. See, for example, Lefever, *Nuclear Arms in the Third World*, pp. 114–117.
72. The following analysis relies partly on interviews with officials, academics, and journalists in Argentina and Brazil in March 1980.
73. "Brazil and Argentina Make It Together," *Latin America Weekly Report*, May 9, 1980, pp. 9–10.
74. John R. Redick, "The Tlatelolco Regime and Non-Proliferation in Latin America" (paper prepared for the World Peace Foundation, March 1980), pp. 10–12, 20–24; "Brazil's Nuclear Dilemma," *Financial Times* (London), March 28, 1980; "Brazil and Argentina to Cooperate," *Nuclear Engineering International*, June 1980, p. 6.
75. Robert M. Levine, "Brazil: Democracy without Adjectives," *Current History*, February 1980, pp. 49–52, 82–83; "The Year in Brazil," *Latinamerican Week*, December 28, 1979, pp. 5–6; "Brazil," *Latin America Regional Reports*, May 30, 1980; "Brazil and Argentina Make It Together," *Latin America Weekly Report*, May 9, 1980, pp. 9–10.
76. See also John R. Redick, "Regional Restraint: U.S. Nuclear Policy and Latin America," *ORBIS* 22, no. 1 (Spring 1978), pp. 193–194.
77. On that goal, see William Perry and Sheila Kern, "The Brazilian Nuclear Program in a Foreign Policy Context," *Comparative Strategy* 1, nos. 1 and 2 (1978), pp. 53–70; H. Jon Rosenbaum, "Brazil's Nuclear Aspirations," in *Nuclear Proliferation and the Near-Nuclear Countries*, ed. Onkar Marwah and Ann Schulz (Cambridge, Mass.: Ballinger, 1975), pp. 53–70.
78. C. H. Waisman, "Incentives for Nuclear Proliferation: The Case of Argentina," in *Nuclear Proliferation and the Near-Nuclear Countries*, ed. Marwah and Schulz, pp. 289–292.
79. *Borba* (Belgrade), December 7, 1975, reprinted in *Survival* 19, no. 3 (May–June 1976), pp. 116–117.
80. "Interview with Col.-Gen. Ivan Kukoc," *Nin* (Belgrade), March 13, 1977, reprinted in *Survival* 20, no. 3 (May–June 1977), pp. 127–129; "Yugoslav Hints Atom Arm Goal," *New York Times*, March 12, 1977.
81. That was the conclusion of persons with whom the author spoke in Belgrade in 1977.
82. For reviews of the overall Soviet position, see Toby Trister Gati, "Soviet Perspectives on Nuclear Proliferation," Discussion paper no. 67 (Santa Monica, California: California Seminar on Arms Control and Foreign Policy, 1975); Gloria Duffy, "Soviet Nuclear Exports," *International Security* 3, no. 6 (Summer 1978); George Quester, *The Politics of Nuclear Proliferation* (Baltimore, Md.: The Johns Hopkins University Press, 1973), pp. 33–55.

83. "Monarch Sees Persian Gulf as Center of Big Power Rivalry," *Journal of Tehran*, September 16, 1975.

84. For an excellent, concise but thorough analysis of the rationale behind the consensus see Herbert Passin, "Nuclear Arms and Japan," in *Asia's Nuclear Future*, ed. Overholt, pp. 67–132. More generally also see Overholt, "Nuclear Proliferation in Eastern Asia," pp. 149–157; John E. Endicott, *Japan's Nuclear Option* (New York: Praeger, 1975); Henry S. Rowen, "Japan and the Future Balance in Asia," *ORBIS* 21, no. 2 (Summer 1977). The following discussion draws on the Passin article as well as on the author's interviews with Japanese officials, academics, and observers in 1977.

85. Michael Pillsbury, "A Japanese Card," *Foreign Policy*, no. 33 (Winter 1978–79), pp. 3–15; Bernard K. Gordon, "Loose Cannon on a Rolling Deck?: Japan's Changing Security Policies," *ORBIS* 22, no. 4 (Winter 1979), pp. 971–974; "Japan's Growing Strategic Role: Soviet Moves Spark Defense Support," *Aviation Week and Space Technology*, January 21, 1980; "Japan's Growing Strategic Role: Broad Military Technology Base Sought," *Aviation Week and Space Technology*, January 28, 1980.

86. This typhoon mentality is also noted with some useful caveats about its uniqueness to Japan by Passin, "Nuclear Arms and Japan," pp. 68–74. See also Hideaki Kase, "Northeast Asian Security: A View from Japan," *Comparative Strategy* 1, nos. 1 and 2 (1978), pp. 100–101.

87. On these constraints and the question of West German nuclear weapons more generally, see Catherine McArdle Kelleher, *Germany and the Politics of Nuclear Weapons* (New York: Columbia University Press, 1975); Horst Menderhausen, "Will West Germany Go Nuclear?" *ORBIS* 16, no. 2 (Summer 1972), pp. 417–427.

88. On the danger of the erosion of the Soviet-American strategic balance, see Secretary of Defense Harold Brown, "Remarks at the Naval War College Convocation," August 20, 1980, p. 2 (photostat); Colin S. Gray, "The Strategic Forces Triad: End of the Road?," *Foreign Affairs* 56, no. 4 (July 1978); "Special Report: Modernizing Strategic Forces," *Aviation Week and Space Technology*, June 16, 1980, passim.

89. Helmut Schmidt, "The 1977 Alastair Buchan Memorial Lecture," *Survival* 20, no. 1 (January–February 1978), pp. 3–5; "A Darkening Sky Over Europe's Grey Areas," *The Economist*, March 31, 1979, p. 33; "NATO Seeks Strategic Counter Force," *Aviation Week and Space Technology*, September 3, 1979, pp. 66–67.

90. "Partager l'arme nucléaire avec les Allemands?: Un entretien avec Georges Buis et Alexandre Sanguinetti," *Le Nouvel Observateur* (Paris), August 20, 1979.

CHAPTER 4

1. For optimistic assessments of the consequences of proliferation, often extrapolated from the experience of the first decades of the nuclear age, see

K. Subrahmanyan, "India: Keeping the Option Open," in *Nuclear Prolifer-ation Phase II*, ed. Robert M. Lawrence and Joel Larus (Lawrence: University Press of Kansas, 1974); Steven J. Rosen, "A Stable System of Mutual Nuclear Deterrence in the Arab-Israeli Conflict," *The American Political Science Review* 71, no. 4 (December 1977), pp. 1367–1383; Paul Jabber, "A Nuclear Middle East: Infrastructure, Likely Military Postures and Prospects for Strategic Stability," ACIS working paper no. 6 (Center for Arms Control and International Security, University of California, Los Angeles, September 1977), pp. 37–39; R. Robert Sandoval, "Consider the Porcupine: Another View of Nuclear Proliferation," *Bulletin of the Atomic Scientists*, May 1976, pp. 17–19; Kenneth N. Waltz, "What Will the Spread of Nuclear Weapons Do to the World?" in *International Political Effects of the Spread of Nuclear Weapons*, ed. John Kerry King (Washington, D.C.: U.S. Government Printing Office, 1979), pp. 165–196; Shai Feldman, "A Nuclear Middle East," *Survival* 23 (May–June 1981), pp. 111–115.

2. Waltz, "What Will the Spread of Nuclear Weapons Do to the World?" pp. 184–190, 194.

3. Ibid., p. 187. For a more general critique of such ethnocentrism, also see Ken Booth, *Strategy and Ethnocentrism* (New York: Holmes & Meier, 1979).

4. See Fred Charles Iklé, "Can Nuclear Deterrence Last Out the Century?" *Foreign Affairs* 51, no. 2 (January 1973), pp. 269–271.

5. See Stockholm International Peace Research Institute, *World Armaments and Disarmament SIPRI Yearbook 1977* (Cambridge: The MIT Press, 1977), pp. 65–67; Joel Larus, *Nuclear Weapons Safety and the Common Defense* (Columbus, Ohio State University Press, 1967), passim.

6. Gavin Kennedy, *The Military in the Third World* (New York: Scribner, 1974), pp. 337–344.

7. The classic study of such strategic interaction is Thomas C. Schelling's "The Reciprocal Fear of Surprise Attack." See Thomas C. Schelling, *The Strategy of Conflict* (New York: Oxford University Press, 1963), pp. 4–22.

8. H. R. Haldeman with Joseph Di Mona, *The Ends of Power* (New York: New York Times Books, 1978), pp. 90–94; Harry G. Gelber, "Nuclear Weapons and Chinese Policy," in *The Superpowers in a Multinuclear World*, ed. Geoffrey Kemp et al. (Lexington, Mass.: Lexington Books, 1974), p. 66.

9. See, among many others, Robert E. Osgood and Robert W. Tucker, *Force, Order, and Justice* (Baltimore: The Johns Hopkins University Press, 1967), pp. 155–156.

10. "India Gives Warning of Atom-Arms Race," *New York Times*, August 16, 1979; and "India, Pakistan Fail to Resolve Key Differences," *Washington Post*, July 17, 1980.

11. Harold Brown, Secretary of Defense, Department of Defense, *Annual Report of Fiscal Year 1981*, pp. 114–117. The discussion here draws on discussions with my colleague George Wittman.

12. Francis P. Hoeber, David B. Kassing, William Schneider, Jr., *Arms, Men,*

and *Military Budgets: Issues for Fiscal Year 1979* (New York: Crane, Russak, 1978), pp. 36–38.

13. For an exaggerated estimate of nuclear weapons' equalizing effect, see Pierre M. Gallois, *The Balance of Terror* (Boston: Houghton Mifflin, 1961).

14. See Klaus Knorr, *On the Uses of Military Power in the Nuclear Age* (Princeton, N. J.: Princeton University Press, 1966).

15. A classic discussion of these risks and uncertainties is Stanley Hoffman, "Nuclear Proliferation and World Politics," in *A World of Nuclear Powers?*, ed. Alastair Buchan (Englewood Cliffs, N.J.: Prentice-Hall, 1966), pp. 96–109.

16. For a generic discussion of the problem see Geoffrey Kemp, *Nuclear Forces for Medium Powers, Part I: Targets and Weapons Systems, Parts II and III: Strategic Requirements and Options*, Adelphi Papers nos. 106 and 107 (London: The International Institute of Strategic Studies, 1974); see also Peter Nailor and Jonathan Alford, *The Future of Britain's Deterrent Force*, Adelphi Paper no. 156 (London: The International Institute for Strategic Studies, 1980).

17. On the significance of this restriction, see Nailor and Alford, *The Future of Britain's Deterrent Force*, pp. 10, 22–23.

18. The most thorough discussion of these issues remains Theodore B. Taylor and Mason Willrich, *Nuclear Theft: Risks and Safeguards* (Cambridge, Mass.: Ballinger, 1974).

19. The widespread civilian reprocessing, transportation, and commercial use of plutonium poses a comparable, if not greater, risk. See ibid.; R. Jeffrey Smith, "Reprocessing May Pose Weapons Threat," *Science* 209, July 11, 1980.

20. Among recent writings on this issue of subnational groups' access to nuclear weapons, see Roberta Wohlstetter, "Terror on a Grand Scale," *Survival* (May–June 1976); Brian Jenkins, "Will Terrorists Go Nuclear?" California Seminar on Arms Control and Foreign Policy, October 1975; David M. Rosenbaum, "Nuclear Terror," *International Security* 1, no. 3 (Winter 1977). For an earlier but very suggestive analysis, also see George H. Quester, "The Politics of Twenty Nuclear Powers," in *The Future of the International Strategic System*, ed. Richard Rosecrance (San Francisco: Chandler, 1972), pp. 66–73.

21. Many of these actions in the Middle East are reviewed in Edward Weisband and Damir Roguly, "Palestinian Terrorism: Violence, Verbal Strategy, and Legitimacy," in *International Terrorism: National, Regional and Global Perspectives*, ed. Yonah Alexander (New York: Praeger, 1976), pp. 258–310.

22. George Quester first pointed to this aspect of the impact of nuclear weapon proliferation on domestic political life. See Quester, "The Politics of Twenty Nuclear Powers," pp. 66–70.

23. For a more detailed discussion of how nuclear weapons could come to play

a role in military coups, with descriptions of past coups, see Lewis A. Dunn, "Military Politics, Nuclear Proliferation, and the 'Nuclear Coup d'Etat,' " *The Journal of Strategic Studies* 1, no. 1 (May 1978), pp. 41–46. See also Edward Luttwak, *Coup d'Etat* (New York: Knopf, 1969); Kennedy, *The Military in the Third World,* passim.

24. D. G. Brennan, "The Risks of Spreading Weapons: A Historical Case," Arms Control and Disarmament, vol. 1 (n.p., 1968), pp. 59–60.

25. See Russell W. Ayres, "Policing Plutonium: The Civil Liberties Fallout," *Harvard Civil Rights–Civil Liberties Review* 10 (1975); Alan F. Westin, "Civil Liberties Implications of U.S. Domestic Safeguards," in Office of Technology Assessment, *Nuclear Proliferation and Safeguards,* app. 3–C, pp. 127–181; Paul Wilkinson, "Terrorism versus Liberal Democracy—The Problems of Response," *Conflict Studies,* no. 67, January 1976; Joseph W. Bishop, "Can Democracy Defend Itself against Terrorism?" *Commentary* 65, no. 5 (May 1978).

26. See especially Ayres, "Policing Plutonium," pp. 413–424.

27. Bishop, "Can Democracy Defend Itself against Terrorism?" p. 58; Irene Dische, "West Germany's War on Terrorism," *Inquiry* 1, no. 16, June 26, 1978, p. 17.

28. Abraham Lincoln, "Special Message to Congress, 1861," in *Abraham Lincoln: Selected Speeches, Messages, and Letters,* ed. T. Harry Williams (New York: Holt, Rinehart, 1957), p. 156.

CHAPTER 5

1. See Lellouche, "How to Break the Rules without Quite Stopping the Bomb: European Views of Carter's Non-Proliferation Policy," and Karl Kaiser, "The Great Nuclear Debate," *Foreign Policy,* no. 30 (Spring 1978), pp. 83–110; Ryukichi Imai, "A Japanese View," in *Nuclear Energy and Nuclear Proliferation: Japanese and American Views,* ed. Ryukichi Imai and Henry S. Rowen (Boulder, Colo.: Westview Press, 1980), part 1.

2. *Nucleonics Week,* January 17, 1980, p. 10; "Reprocessing Company's President Named," *Nuclear Engineering International,* January 1980, p. 7; "Survey of Japan," *Nuclear Engineering International,* December 1979, pp. 53–60, 66–70, 72–74.

3. *Nucleonics Week,* November 2, 1978, p. 10; and October 4, 1979, pp. 1–2; "New Sites for Reprocessing," *Nuclear Engineering International,* November 1980, p. 9.

4. *Nucleonics Week,* December 4, 1980, pp. 5–6; "France Forging Ahead on Fast Reactors," *Nuclear News,* November 1980, pp. 99–104.

5. *Nucleonics Week,* April 21, 1977, pp. 1–2; July 3, 1980, p. 4; and July 10, 1980, p. 2.

6. "Uncertainty Hovers over the Uranium Market," *Nuclear Engineering International,* October 1980, pp. 28–30; International Nuclear Fuel Cycle Evaluation, *Reprocessing, Plutonium Handling, Recycle,* Report of INFCE Work-

ing Group 4 (Vienna, Austria: International Atomic Energy Agency, 1980), pp. 8–9; *Nucleonics Week,* December 4, 1980, pp. 5–6.

7. Dr. P. Bauder, A. Horncastle, G. Lurf, and Dr. M. Stephany, "Competing for the Non-USA Enrichment Markets," *Nuclear Engineering International,* October 1980, pp. 37–41.

8. *Nucleonics Week,* December 4, 1980, pp. 5–6; "France Forging Ahead on Fast Reactors," *Nuclear News,* November 1980, pp. 99–104.

9. "Reprocessing Company's President Named," *Nuclear Engineering International,* January 1980, p. 7; *Nucleonics Week,* January 17, 1980, p. 10.

10. *Nucleonics Week,* December 21, 1978, p. 10; October 4, 1979, pp. 1–2; and January 17, 1980, p. 10; "Local Government Delays Construction," *Nuclear Engineering International,* November 1980, p. 10.

11. The following analysis draws partly on Paul Bracken, "Fuel Assurances in Proliferation Control Strategies," (Hudson Institute: HI–3094/2–P, January 10, 1979).

12. "To Reprocess Better than to Store," *Nuclear Engineering International,* November 1980, p. 4.

13. *Nucleonics Week,* September 11, 1980, p. 1.

14. Foreign Assistance Act of 1961, Sections 669 and 670.

15. Export-Import Bank Act, Section 2(b) (4).

16. Foreign Assistance Act of 1961, Sections 669 and 670; Atomic Energy Act of 1954, as amended by the Nuclear Non-Proliferation Act of 1978, Section 129.

17. Statute of the International Atomic Energy Agency (as amended up to June 1, 1973), Article XII C.

18. Guideline 14(c), reprinted in Atlantic Council, *Nuclear Power and Nuclear Weapons Proliferation,* vol. 2, app. D (Boulder, Colo.: Westview Press, 1978), p. 66.

19. "Resource-short Taiwan Plunges Ahead with Nuclear Power Plans," *Christian Science Monitor,* August 12, 1980; "South Korea's Nuclear Program," *Financial Times* (London), May 7, 1980; International Nuclear Fuel Cycle Evaluation, *Fuel and Heavy Water Availability,* Report of INFCE Working Group 1 (Vienna: IAEA, 1980), pp. 46–47.

20. International Monetary Fund, *Direction of Trade Yearbook, 1980,* (Washington, D.C., International Monetary Fund, 1980), pp. 229–230.

21. Robert S. Jaster, *South Africa's Narrowing Security Options,* Adelphi Papers no. 159 (London: International Institute for Strategic Studies, 1980), p. 40.

22. "Political Turmoil Reported to Hurt Libya Economy," *New York Times,* June 27, 1980.

23. *World Bank, 1979 Annual Report,* pp. 52–53.

24. "The U.S. Capitulates on Debt Relief," *World Business Weekly,* June 30, 1980; "Brazil's Foreign Debt: Haggling over $12 Billion," *World Business Weekly,* March 17, 1980; "Banks Trim Loans to Third World Amid Fears of Repayment Problem," *New York Times,* April 14, 1980.

25. The World Bank, *World Development Report, 1979* (Washington, D.C.: The World Bank, August 1979), pp. 28–32, 94–98; "Korea Gets New Loans for Growth," *New York Times*, April 14, 1980; "Outlook for Asia's Star Performers," *Asian Wall Street Journal*, February 25, 1980.

26. Bank for International Settlements, "Maturity Distribution of International Bank Lending—end–June 1980," December 1980, see table following p. 4 (mimeo).

27. "Investigation of Korean-American Relations," pp. 79–80; "U.S. Kept S. Korea from Making A-Arms," *International Herald Tribune* (London), November 6, 1978; "South Korea Drops Plan to Buy a Nuclear Plant from France," *New York Times*, January 30, 1976; "Seoul Officials Say Strong U.S. Pressure Forced Cancellation of Plans to Purchase a French Nuclear Plant," February 1, 1976.

28. Ernest W. Lefever, *Nuclear Arms in the Third World: U.S. Policy Dilemma* (Washington, D.C.: The Brookings Institution, 1979), p. 89 "Taipei Treads Lightly," *Far Eastern Economic Review*, January 6, 1978; "Taiwan Forces Reportedly buy Israeli Missiles," *New York Times*, April 6, 1977; "Reprocessing to Halt, Administration Says," *Nuclear News*, November 1976; "East, West Anxiously Watch Taiwan's Nuclear Progress," *International Herald Tribune* (London), March 2, 1977.

    However, if the United States continues to reduce its ties to Taiwan and fails to provide conventional arms, a future threat of sanctions might not suffice to counterbalance Taiwan's increased incentives to acquire the bomb for security reasons.

29. "U.S.-Soviet Exchange about South Africa Said to Improve Ties," *New York Times*, August 29, 1977.

30. Ibid.; "United States Asks South Africa about A-Bomb Plan," *New York Times*, August 21, 1977; "South Africa Stirs New A-Arms Flurry," *New York Times*, August 31, 1977.

31. "President Carter's News Conference of August 23," *Department of State Bulletin*, September 19, 1977, p. 376.

32. Margaret Doxey, *Economic Sanctions and International Enforcement* (London: Oxford University Press, 1971), pp. 51–57; George W. Baer, "Sanctions and Security: The League of Nations and the Italian-Ethiopian War, 1935–1936," *International Organization* 27, no. 2 (Spring 1973), p. 179.

33. Harry R. Strack, *Sanctions: The Case of Rhodesia* (Syracuse; N.Y.: Syracuse University Press, 1978), pp. 237–238, 241–244; Doxey, *Economic Sanctions*, pp. 75–81; Donald L. Losman, *International Economic Sanctions: The Cases of Cuba, Israel, and Rhodesia* (Albuquerque: University of New Mexico Press, 1979), pp. 92–121.

34. Anna P. Schreiber, "Economic Coercion as an Instrument of Foreign Policy: U.S. Economic Measures against Cuba and the Dominican Republic," *World Politics* 25 (April 1973), pp. 394–405; Losman, *International Economic Sanctions*, pp. 42–46.

35. "Iran's Islamic Revolution Lurches Out of Control," *World Business Weekly*, July 28, 1980; "As Iran's Fiscal Position Degenerates, Assets Become U.S.

NOTES 201

Bargaining Chip," *Wall Street Journal,* July 8, 1980; Terrence Smith, "Putting the Hostages First," *New York Times Magazine,* May 17, 1981, passim.
36. See Klaus Knorr, "International Economic Leverage and Its Uses," in *Economic Issues and National Security,* ed. Klaus Knorr and Frank N. Trager (Lawrence, Kansas: Regents Press of Kansas, 1977), pp. 106–109.
37. Judith Miller, "When Sanctions Worked," *Foreign Policy,* no. 39 (Summer 1980), p. 125.
38. Schreiber, "Economic Coercion as an Instrument of Foreign Policy," p. 405.
39. Atlantic Council, *Nuclear Power and Nuclear Weapons Proliferation,* vol. 2, app. D, p. 66.
40. Secretary of State Cyrus Vance, *The Department of State Bulletin,* August 1978.
41. See J. J. Martin, "Nuclear Weapons in NATO's Deterrent Strategy," *ORBIS* 22, no. 4 (Winter 1979).
42. Leslie H. Brown, *American Security Policy in Asia,* Adelphi Papers No. 132 (London: International Institute for Strategic Studies, 1977), pp. 31–32.
43. Well-argued representatives of this genre are William Epstein, *The Last Chance* (New York: The Free Press, 1976), esp. chap. 14; SIPRI, *Postures for Non-Proliferation: Arms Limitation and Security Policies to Minimize Nuclear Proliferation* (New York: Crane, Russak, 1979), pp. 39–66 (SIPRI Study by Enid Schoettle).
44. Subrahmanyan, "India: Keeping the Option Open," in *Nuclear Proliferation Phase II,* ed. Robert M. Lawrence and Joel Larus (Lawrence: University Press of Kansas, 1974), pp. 123–124; George Quester, *The Politics of Nuclear Proliferation* (Baltimore: The Johns Hopkins University Press, 1973), pp. 70–71.
45. Steven J. Rosen, "Nuclearization and Stability in the Middle East," in *Nuclear Proliferation and the Near-Nuclear Countries,* ed. Onkar Marwah and Ann Schulz (Cambridge, Mass.: Ballinger, 1975), pp. 166ff.
46. *Borba,* December 7, 1975, reprinted as "Yugoslavia and Nuclear Weapons," in *Survival* 18 (May–June 1976), pp. 116–117.
47. See Shelton L. Williams, *The U.S., India, and the Bomb* (Baltimore: The Johns Hopkins University Press, 1969); Subrahmanyan, "India: Keeping the Option Open," passim.
48. Rising West German and European concern that their interests not be sacrificed in SALT has been evident in recent years. See Kurt Birrenbach, "European Security: NATO, SALT and Equilibrium," *ORBIS* 22, no. 2 (Summer 1978); Laurence Martin, "SALT and U.S. Policy," *The Washington Quarterly* 2, no. 1 (Winter 1979), pp. 31–34; Lothar Ruehl, "NATO Europeans Call for a Say in the Drafting of SALT III," *The Atlantic Community Quarterly* 16, no. 1 (Spring 1978).
49. Donald G. Brennan, "A Comprehensive Test Ban: Everybody or Nobody," *International Security* 1, no. 1 (Summer 1976), pp. 92–117.
50. "Indian Officials Hail Carter's Visit as a Great Diplomatic Success," *New York Times,* January 4, 1978.

51. "Africa's Horn: New Alliances," *New York Times,* July 28, 1977; "U.S. Is Adopting a Neutral Policy in Africa's Horn," *New York Times,* October 3, 1977; "U.S. 'Countermeasures' on Horn?" *Christian Science Monitor,* February 27, 1978.

52. "Dropping Shah Risks Offending a King," *Christian Science Monitor,* January 11, 1979; "After the Shah—a New Saudi Diplomacy?" *Christian Science Monitor,* January 18, 1979.

53. See Shirin Tahir-Kheli, "Pakistan's Nuclear Option and U.S. Policy," *ORBIS* 22, no. 2 (Summer 1978), p. 361.

54. For helpful discussion of this issue, see Richard Burt, "Nuclear Proliferation and Conventional Arms Transfers: The Missing Link," California Seminar on Arms Control and Foreign Policy, September 1977.

55. John J..Mearsheimer, "Precision-guided Munitions and Conventional Deterrence," *Survival* 21, no. 2 (March–April 1979).

56. David Ronfeldt and Caesar Sereseres, "U.S. Arms Transfers, Diplomacy, and Security in Latin America and Beyond," RAND Corporation, P-6005, October 1977, p. 39.

57. Some of the ideas in the next paragraphs are developed at greater length in Lewis A. Dunn, "Some Reflections on the 'Dove's Dilemma'," *International Organization* 35, no. 1 (Winter 1981), pp. 181–192.

58. On NFZs, see Epstein, *The Last Chance,* pp. 207–220; Alfonso Garcia Robles, "The Latin American Nuclear-Weapon-Free Zone," Stanley Foundation Occasional Paper 19, May 1979.

CHAPTER 6

1. See Ryukichi Imai, "Non-proliferation: A Japanese Point of View," *Survival* 21, no. 2 (March–April 1979), p. 56.

2. "Big Powers Urged to Check Nuclear Arms Spread," *Washington Post,* August 13, 1980.

3. "Israel Angered as French Send Uranium to Iraq," *Washington Post,* July 20, 1980.

CHAPTER 7

1. The following draws on discussions with my former colleague at the Hudson Institute, William Schneider, Jr.

2. On command-and-control concepts, see Joel Larus, *Nuclear Weapon Safety and the Common Defense* (Columbus: Ohio State University Press, 1967), chap. 11.

3. For a proposal to earmark ten to twenty obsolete *Minuteman* ICBMs for this purpose or 100 new small ICBMs in the future, see Richard L. Garwin, "Reducing Dependence on Nuclear Weapons: A Second Nuclear Regime," in David C. Gompert et al., *Nuclear Weapons and World Politics: Alternatives for the Future* (New York: McGraw-Hill, 1977), p. 132.

4. For a discussion of the limits, opportunities, and approaches to conventional arms control, see Yair Evron, *The Role of Arms Control in the Middle East,* Adelphi Papers No. 138 (London: The International Institute for Strategic Studies, 1977).

5. See also Evron, *The Role of Arms Control in the Middle East,* pp. 17–19; Thomas C. Schelling, *The Strategy of Conflict* (New York: Oxford University Press, 1963), pp. 242–251; Jonathan Alford, ed., *The Future of Arms Control: Part III Confidence-Building Measures,* Adelphi Papers No. 149 (London: The International Institute for Strategic Studies, 1979), esp. pp. 4–13, 23–29.

6. Alford, *Confidence-Building Measures in Europe,* p. 12.

7. The following draws heavily on discussions with my former colleague at Hudson Institute, William Schneider, Jr.

8. On the limits to transformation of international politics, see Stanley Hoffmann, *Primacy or World Order: American Foreign Policy since the Cold War* (New York: McGraw-Hill, 1978), pp. 161–182; Raymond Aron, *Peace and War* (Garden City, N.Y.: Doubleday, 1966), esp. part 4.

9. Herman Kahn, "Nuclear Proliferation and Rules of Retaliation," *Yale Law Journal* 76, no. 1 (November 1966); Garwin, "Reducing Dependence on Nuclear Weapons," pp. 130–132.; Alton Frye, "How to Ban the Bomb: Sell It," *New York Times Magazine,* January 11, 1976; Richard Ullman, "No first Use of Nuclear Weapons," *Foreign Affairs* 50, no. 4 (July 1972).

10. The best critique of such collective security schemes remains Inis L. Claude, *Power and International Relations* (New York: Random House, 1962), pp. 150–204.

11. J. J. Martin, "Nuclear Weapons in NATO's Deterrent Strategy," *ORBIS,* 22, no. 4 (Winter 1979).

12. See Herman Kahn, *On Escalation* (New York: Praeger, 1965), chap. 6.

INDEX

ABM. *See* Treaty on the Limitation of Anti-Ballistic Missile Systems

Accidental detonation. *See* Nuclear weapons, accidental detonation of

Africa, 12, 56, 126, 138

Algeria, 91–92

Alliances and security guarantees, 125–28, 138, 141, 153–57. *See also individual countries*

Arab-Israeli conflict, 71, 83, 87–88, 126, 128, 132, 138. *See also* Palestine Liberation Organization

Argentina, 78; nuclear programs of, 15, 25, 35–36, 39, 59, 60, 61, 108, 125; as a nuclear supplier, 41, 42, 43

Arms control, 157–59, 175. *See also* Nonproliferation policies

Arms transfers, 128–30. *See also individual countries*

Asia, 11, 12, 58, 110

Atlantic Alliance, 9, 12. *See also* North Atlantic Treaty Organization

Atomic bomb, 28; use of, on Japan, 2, 6, 15

Atomic bomb building, 5, 14–15. *See also* Manhattan Project; Nuclear weapons proliferation

Australia, 100

Baruch Plan, 6–7

Begin, Menachem, 52

Berlin Crises, 19

Black marketing, 37, 39–41, 92, 103, 146, 177

Brazil, 59–60, 78, 109; nuclear programs of, 25, 33, 39, 60–61, 108, 125; as a nuclear supplier, 41–43, 50

Britain. *See* United Kingdom

Canada, 11–12, 36, 57

Carter, Jimmy, 33, 34, 53, 99, 126

Carter Administration, 25–26, 35, 57, 96, 97

China, 77, 124, 169, 171 (*see also* Korean War); nuclear weapons acquisition by, 1, 10–11; and India, 48, 132; and Taiwan, 56, 57–58, 78

Civil liberties, 92–94, 166–67

Comprehensive test ban treaty (CTB), 121, 123, 124–25

Conflict-prone regions, 82; nuclear weapons proliferation in, 2, 75, 76, 95, 149, 162, 176

Conventional arms, 49, 57, 77, 128–30, 140, 152. *See also* Nuclear weapons

Conventional forces, 11, 48, 69, 123, 159. *See also* Military intervention; United States, armed forces abroad

Costs of nuclear weapons acquisition, 10, 60, 64, 65, 102, 110, 116, 123,

Costs of nuclear weapons *(continued)* 135. *See also* Political disincentives to and costs of acquisition
CTB. *See* Comprehensive test ban treaty
Cuba, 113, 114
Cuban Missile Crisis, *1962*, 19, 20, 23, 77

De Gaulle, Charles, 9, 17
Desai, Morarji, 47, 124
Deterrence, 21, 48, 49, 55, 69, 70, 161, 165. *See also* First-use ban; North Atlantic Treaty Organization, deterrence policy of; Second-strike capability
Developing countries. *See* Third World
Disincentives to nuclear weapons acquisition, 13–14, 17–18. *See also nuclear programs of individual countries*
Dissidents. *See* Separatist movements

Economic embargo, 55, 109, 112, 115. *See also* Sanctions
Egypt, 49, 53, 70, 110, 147
Eisenhower Administration, 19
Europe, Eastern, 19, 62, 67. *See also* Warsaw Pact
Europe, Western, 18, 87–88, 101, 179. *See also* Atlantic Alliance; North Atlantic Treaty Organization

Fail-safe procedures, 22, 23, 74
Federal Republic of Germany. *See* Germany (Federal Republic)
Figueiredo, João, 59, 60
First-use ban, 123, 158, 168–73, 174, 175
Ford, Gerald, 33
Foreign Assistance Act, *1961*, 104, 105
France, 67, 80, 87, 88, 102; nuclear weapons acquisition by, 1, 8–10,

17; as a nuclear supplier, 27n, 33, 36, 37, 44–45, 50, 98

Germany (Federal Republic), 66–67, 80, 87, 88, 93 *(see also* Berlin Crises)*; nonnuclear status decision of, 11, 12, 16, 17, 98, 123, 126; as a nuclear supplier, 33, 35–36, 59
Gray marketing, 37–39, 103, 142, 146, 177, 178

Hussein, Saddam, 50, 51

IAEA. *See* International Atomic Energy Agency
Incentives for nuclear weapons acquisition, 16–17, 118. *See also* Political incentives for acquisition; Security incentives for acquisition
India, 77, 109, 124–25, 130, 137 *(see also* China, and India)*; nuclear explosives activities of, 11, 17, 25, 44–48, 112, 115–16, 127; as a nuclear supplier, 39, 41, 42, 43; and Pakistan, 44–48, 70–71, 77, 124–25, 127, 137, 139
INFCE. *See* International Nuclear Fuel Cycle Evaluation
Intelligence requirements and activities, 53–54, 90, 103, 111, 135–36, 137, 146–47, 160–61; American, 2, 40, 48, 165, 166
International Atomic Energy Agency (IAEA), 35, 105, 107, 111, 143, 144
International Nuclear Fuel Cycle Evaluation (INFCE), 34–35
Iran, 63–64, 113. *See also* Iraq-Iran war
Iraq, 63, 109, 125; nuclear explosives activities of, 27n, 31, 36, 41, 42–43, 50–53; nuclear reactor bombing by Israel, 31, 50, 77, 147
Iraq-Iran war, 70, 77, 113, 126
Israel, 38–39, 83, 109, 110, 112, 130 *(see also* Arab-Israeli conflict; Iraq,

nuclear reactor bombing by Israel);
nuclear weapons acquisition by, 25,
40, 48–49, 106, 122, 125
Italy, as a nuclear supplier, 26, 36,
50

Japan, 64–66, 67, 80, 126, 179 (*see
also* Atomic bomb, use of, on
Japan); nonnuclear status decision
of, 11, 12–13, 16, 17, 59, 123;
nuclear programs of, 96, 98, 99,
101

Kahn, Herman, 168–69, 172*n*
Kennedy, John, 11, 20, 23
Khomeini, Ruhollah, 63, 64, 113
Khrushchev, Nikita, 20, 23
Korea, South, 119, 125–26, 128,
133; nuclear explosives activities
of, 25, 57–59, 99, 108, 110–11
Korean War, 10, 19, 20, 23, 121

Latin America, 41, 59–61
Libya, 70, 109, 125; nuclear
explosives activities of, 15, 31, 36,
42, 51–53
London Nuclear Suppliers Group,
34, 35, 36, 41, 42, 105, 107, 117,
177

Manhattan Project, 1, 5–6, 7–8, 12,
25
Middle East, 31, 48–53, 70, 77, 127,
128. *See also* Arab-Israeli conflict
Military intervention, 81–83, 126. *See
also* Risks of military intervention;
United States, armed forces abroad
Missiles, 29–30, 39, 66, 70, 154, 176.
See also Treaty on the Limitation of
Anti-Ballistic Missile Systems
Mitterand, François, 36, 51, 102, 131

Nagasaki bombing. *See* Atomic bomb,
use of, on Japan
NATO. *See* North Atlantic Treaty
Organization

NFZ. *See* Nuclear free zones
Nigeria, 55–56, 127
Nixon Administration, 57
Nonproliferation policies, 1, 2–3, 33,
95, 133, 167, 177–79. *See also*
Alliances and security guarantees;
Arms transfers; Carter
Administration; Comprehensive
test ban treaty; Nuclear export
controls; Nuclear free zones;
Nuclear Non-Proliferation Act;
Safeguards; Sanctions; Treaty of
Tlatelolco; Treaty on the
Limitation of Anti-Ballistic Missile
Systems; Treaty on the
Nonproliferation of Nuclear
Weapons
North Atlantic Treaty Organization
(NATO), 12, 66–67, 88, 120, 123,
126, 158*n*, 179; deterrence policy
of, 13, 14, 80, 119, 168, 172, 174
NPT. *See* Treaty on the
Nonproliferation of Nuclear
Weapons
Nuclear blackmail, 51–52, 77, 83,
87–88, 127, 138, 155
Nuclear energy, 18, 96–102, 108,
177. *See also nuclear programs of
individual countries;* Nuclear
reactors
Nuclear export controls, 2, 102–04,
178, 179
Nuclear free zones (NFZ), 131–32,
177
Nuclear fuel, 25, 34–35, 96, 97–102,
111
Nuclear Non-Proliferation Act, *1978*,
35, 104–05, 116
Nuclear peace and war, 18–23,
75–78
Nuclear reactors, 8, 10, 14, 32–33,
34, 96, 97. *See also* Nuclear energy
Nuclear taboo, 17–20
Nuclear tests, 2, 44, 138–42, 176. *See
also nuclear explosives activities,
nuclear programs, nuclear weapons*

Nuclear tests *(continued)*
  acquisition of individual countries;
  Peaceful nuclear explosive
Nuclear weapons, 32 *(see also* Atomic
  bomb; Fail-safe procedures;
  Missiles; Threat of nuclear attack);
  accidental detonation of, 22,
  71–73, 75, 150, 151–52; delivery
  of, 28–31, 49, 73, 154, 166;
  unauthorized use by military of,
  73–75, 76, 151
Nuclear weapons proliferation, 1,
  134, 176–77, 180. *See also nuclear
  explosives activities and nuclear
  programs of individual countries;*
  Atomic bomb building; Conflict-
  prone regions, nuclear weapons
  proliferation in; Nonproliferation
  policies; Risks of nuclear weapons
  acquisition

Pakistan, 105, 109, 113, 114 *(see also*
  India, and Pakistan); nuclear
  explosives activities of, 11, 27, 31,
  33, 38, 44–48, 77, 103, 108; as a
  nuclear supplier, 41, 43, 51
Palestine Liberation Organization
  (PLO), 89–90, 126
Peaceful nuclear explosive (PNE), 17,
  46, 112, 139
Perón, Juan, 15
PLO. *See* Palestine Liberation
  Organization
PNE. *See* Peaceful nuclear explosive
Political disincentives to and costs of
  acquisition, 17–18, 47, 83, 107,
  140, 148, 161, 171, 179
Political incentives for acquisition, 1,
  6, 7, 8, 11, 16, 125, 137

Qaddafi, Muammar, 15, 31, 51, 52

Reagan Administration, 36, 58, 98,
  102, 105, 114, 121, 127

Reprocessing of spent fuel, 25, 96,
  111
Rhodesia, 112–13
Richter, Ronald, 15, 39
Risks of military intervention, 85–87,
  128, 129, 155, 162–64
Risks of nuclear weapons acquisition,
  102, 107, 114–17, 140. *See also*
  Threat of nuclear attack
Risks of sanctions, 114–17, 122, 135,
  148

Safeguards, 17–18, 142–44
Sanctions, 104–21 passim, 136, 137,
  141, 143–44, 179. *See also*
  Economic embargo; Risks of
  sanctions
Saudi Arabia, 31, 49, 87, 126, 129
Second-strike capability, 21–22
Security incentives for acquisition, 1,
  8, 16, 45–46, 79, 53, 122
Separatist movements, nuclear threat
  by, 74, 90, 91–92
South Africa, 43, 83, 109, 115, 125;
  nuclear weapons acquisition by, 27,
  38–39, 53–56, 106, 111–12, 127
South Korea. *See* Korea, South
Soviet Union, 77, 174, 179; nuclear
  weapons acquisition by, 1, 6–7; as
  a nuclear supplier, 10, 36, 51;
  threat of nuclear attack on, 62, 64,
  78, 80, 82–84, 170–72, 176; and
  the United States, 79–81, 85–87,
  162–64, 174 *(see also* Berlin Crises;
  Cuban Missile Crisis; Military
  intervention)
Stalin, Joseph, 7
Suppliers of nuclear technology, 2,
  36, 41–43, 97. *See also individual
  countries as nuclear suppliers;* Black
  marketing; Gray marketing;
  London Nuclear Suppliers Group
Sweden, 13, 17, 39
Switzerland, 13–14, 35–36

Taiwan, 109, 125, 127, 128–29 (*see also* China, and Taiwan); nuclear explosives activities and programs of, 25, 39, 56–59, 108, 111
Technical assistance, 10, 150–52, 159
Technical constraints, 16, 24–28, 31–32, 43, 56, 61, 73, 102
Terrorists, 93 (*see also* Palestine Liberation Organization); nuclear threat by, 3, 30–31, 74, 88–90, 92, 159, 164–65, 168, 173
Third World, 3, 96–97, 117, 160, 166, 171, 179; nuclear programs of, 26, 41, 99–100
Threat of nuclear attack, 78, 79–81, 119–21, 159, 168, 172–73, 180. *See also* First-use ban; Risks of military intervention; Second-strike capability; Separatist movements, nuclear threat by; Soviet Union, threat of nuclear attack on; Terrorists, nuclear threat by; United States, threat of nuclear attack on
Tito, 14
Treaty of Tlatelolco, 60, 131
Treaty on the Limitation of Anti-Ballistic Missile Systems, *1972* (ABM), 80, 88
Treaty on the Nonproliferation of Nuclear Weapons (NPT), 17–18, 42, 61, 103, 119, 122, 142–43; nonsignatories of, 37, 42, 60, 130; withdrawal of signatories of, 100, 106, 128, 134, 144–46

United Kingdom, 87, 88, 93, 102; nuclear weapons acquisition by, 1, 7–8
United States, 93, 94, 99–102, 137, 149, 152–53, 178–79 (*see also* Alliances and Security guarantees; Berlin Crises; Cuban Missile Crisis; Intelligence requirements and activities, American; Korean War; Manhattan Project; Soviet Union, and the United States); policies toward allies of, 3, 12, 68, 125–28, 140, 153–57 (*see also* Atlantic Alliance; North Atlantic Treaty Organization); threat of nuclear attack on, 9, 80, 82–84, 119, 164–66, 173, 176; armed forces abroad, 57, 58, 82, 160, 161–62 (*see also* Military intervention)

Vorster, B. J., 53–54

Warsaw Pact, 119, 120, 172
West Germany. *See* Germany (Federal Republic)
World War II. *See* Atomic bomb, use of, on Japan; Manhattan Project

Yugoslavia, 14, 25, 61–63, 78, 83, 122

Zia ul Haq, Mohammed, 44, 46, 139
Zimbabwe. *See* Rhodesia